W9-CXS-905

PROTECTION OF METALS FROM CORROSION IN STORAGE AND TRANSIT

ELLIS HORWOOD SERIES IN CORROSION AND ITS PREVENTION

Series Editors:
A. D. MERCER, Principal Scientific Officer, National Physical Laboratory and Scientific Secretary, European Federation of Corrosion
Dr. D. R. HOLMES, Manager/Consultant, High Temperature Corrosion Projects, National Corrosion Service, National Physical Laboratory
Dr. D. H. SHARPE, O.B.E., lately General Secretary, Society of Chemical Industry; formerly General Secretary, Institution of Chemical Engineers; and former Technical Director, Confederation of British Industry.

This new series provides up-to-date compelling texts on the theory and practice of corrosion science and technology. There is special emphasis on cost-effective preventative measures and on the most recent developments in new materials and techniques for the study of corrosion products. The books will usually be at the post-graduate research or practicing corrosion engineer technologist level, but will also be suitable for informing advanced undergraduate and master courses of the most recent developments. Together, they will provide substantial coverage of the latest developments in corrosion phenomena and protective measures.

PROTECTION OF METALS FROM CORROSION IN STORAGE AND TRANSIT
P. D. DONOVAN, Principal Scientific Officer, Ministry of Defence, Royal Armament Research and Development Establishment
HIGH ALLOY STAINLESS STEELS FOR CRITICAL SEAWATER APPLICATIONS
Editors: J. W. OLDFIELD and P. GUHA

Books in Corrosion Science Published for the Institution of Corrosion Science and Technology

CATHODIC PROTECTION: Theory and Practice
Editors: V. ASHWORTH, Global Corrosion Consultants and C. J. L. BOOKER, City of London Polytechnic
DEWPOINT CORROSION
Editor: D. R. HOLMES, Manager/Consultant, High Temperature Corrosion Projects, National Corrosion Service, National Physical Laboratory
COATINGS AND SURFACE TREATMENT FOR CORROSION AND WEAR RESISTANCE
Editors: K. N. STRAFFORD and P. K. DATTA, School of Materials Engineering, Newcaste upon Tyne Polytechnic, and C. G. GOOGAN, Global Corrosion Consultants Limited, Telford
PREDICTION OF MATERIALS PERFORMANCE IN PLANTS OPERATING WITH CORROSIVE ENVIRONMENTS
Editors: J. E. STRUTT and J. R. NICHOLLS, Cranfield Institute of Technology, Bedford

PROTECTION OF METALS FROM CORROSION IN STORAGE AND TRANSIT

P. D. DONOVAN
Principal Scientific Officer
Ministry of Defence
Royal Armament Research and Development Establishment

ELLIS HORWOOD LIMITED
Publishers · Chichester

Halsted Press: a division of
JOHN WILEY & SONS
New York · Chichester · Brisbane · Toronto

First published in 1986 by
ELLIS HORWOOD LIMITED
Market Cross House, Cooper Street, Chichester, West Sussex, PO19 1EB, England

The publisher's colophon is reproduced from James Gillison's drawing of the ancient Market Cross, Chichester.

Distributors:
Australia and New Zealand:
Jacaranda-Wiley Ltd., Jacaranda Press,
JOHN WILEY & SONS INC.
GPO Box 859, Brisbane, Queensland 4001, Australia
Canada:
JOHN WILEY & SONS CANADA LIMITED
22 Worcester Road, Rexdale, Ontario, Canada
Europe and Africa:
JOHN WILEY & SONS LIMITED
Baffins Lane, Chichester, West Sussex, England
North and South America and the rest of the world:
Halsted Press: a division of
JOHN WILEY & SONS
605 Third Avenue, New York, NY 10158, USA

British Library Cataloguing in Publication Data
Donovan, P.D.
Protection of metals from corrosion in storage and transit. —
(Ellis Horwood series in corrosion and its prevention)
1. Corrosion and anti-corrosives
I. Title
620.1′623 TA462

Library of Congress Card No. 86–7185

ISBN 0–85312–690–9 (Ellis Horwood Limited)
ISBN 0–470–20332–3 (Halsted Press)

Phototypeset in Times by Ellis Horwood Limited
Printed in Great Britain by The Camelot Press, Southampton

Table of Contents

Preface

The scope of this book reflects my own experience in carrying out research into corrosion mechanisms, assessing the susceptibility to corrosion of alloys, evaluating the protection offered by metal finishes and exploring the properties of new materials, and in applying the results to the design of new equipments to ensure that they did not suffer corrosion in manufacture, transit, storage or use, and in investigating instances of corrosion when they have occurred.

During the course of my career many changes in materials, transport, storage conditions and manufacturing methods have taken place. The intensive capitalisation of processing and the application of management techniques to the analysis of designs for cost effectiveness (or Value Engineering), reliability, availability, maintainability, durability and safety, now more than ever reveal the key role of corrosion and its prevention. Nevertheless corrosion still strikes unexpectedly, as quixotic as the Scarlet Pimpernel, mocking those who predict reliabilities through elaborate fault analyses to fine levels of precision. The cause is typically an oversight of some apparently minor change of materials, processing or storage conditions, or ignorance of the potential problems of incompatibility between some of the materials used.

In this book I have tried to present the problems of corrosion and the principles of protection in a way that is not essentially linked to a particular type of design or engineering discipline but as a general philosophy which may be applied to all designs. I have attempted to make the text generally intelligible to those with some technical training without compromising the chemical explanation. I have followed the approach which has proved successful within the Ministry of Defence in guiding designers in guarding against corrosion, and in training specialists in Quality Assurance in the techniques of metal finishing. Once a basic understanding of a comparatively few principles has been attained their general application to what appears to be a widely varied range of phenomena provides an intellectually satisfying application of theory to practice. It was this later connection which inspired U. R. Evans to speak of the sheer joy of seeing the beauty of the scientific pattern emerge from a study of the otherwise apparently ugly and morbid subject of corrosion. I have also tried to make the text useful to corrosion specialists by providing a comprehensive account of the corrosion hazards encountered in storage and a summary of the methods of protection available.

My own work on corrosion has been as a member of a team, and part of the wider community of those involved in corrosion research throughout the country which has provided me with both technical support and lasting friendship. I would like to thank those who have been colleagues during my studies of corrosion, especially Dr E. Longhurst, Jack Andrew, John Stringer, Tom Heron and Henry Cole, and the many members of committees on which I have served, for their work and continued help and inspiration on which much of this book depends. In preparing the text I have been fortunate in the forebearance of my family and the help of my wife Angela and daughter Lucy in typing and preparing illustrations. I am very grateful to Henry Cole, Derek Holmes and John Stringer for their dedication in reading the manuscript and for their many suggestions for improving the text. Any remaining deficiencies or mistakes are mine.

1

The Need

Metals have shaped the development of the modern world and are now fundamental to our material well-being. Goods made in whole or in part of metals are a major portion of the world's increasing wealth and the distribution of these ever more sophisticated artefacts within and across national boundaries is a characteristic of our civilisation.

As the wealth, aspirations and technological capabilities of nations have risen, the resulting pattern of exchanges of manufactured products has grown in both extent and complexity. An informal system has evolved for moving large quantities of goods rapidly and efficiently around the globe in response to market demands. This system, based on bulk transport, mechanical handling, large interstage storage areas and fast communications, has an inherent flexibility which allows commodities to be rerouted to meet changes in established markets or to develop new markets.

This capability in transport, communications and marketing has to be matched by an ability of the supporting technologies to ensure that goods remain fit for their intended purpose and, just as importantly, are of 'marketable' quality when they are delivered into the hands of the customer. To the customer this means that they should not only be in a serviceable condition but that they should also look attractive and be 'as new'. If merchandise is to arrive in this pristine state it must be protected not only from physical damage but also from the more insidious depredations of staining, rotting, attack by animals, chemical interactions and corrosion.

From the earliest days of trading, protective methods evolved which were often effective, although some spoilage was usually considered inevitable. A pragmatic and evolutionary approach often proved adequate for the comparatively few types of items of simple and robust design which then moved predictably over the same routes, with any changes to the pattern of trade, or the types of commodity transported, occurring only slowly. Methods of protection could evolve on a basis of trial and error, to meet new hazards which resulted from modifications in designs or conditions of transport. However, a more fundamental understanding was needed to combat the sudden onset of serious deterioration when rapid changes in the designs of merchandise, or in the destinations to which they were being sent, became common practice.

The need for greater understanding was very forcibly presented to the United Kingdom during the second world war when the products of the new technologies such as radio sets, aircraft, vehicles, high performance guns, ammunition and spare components were frequently found to be severely corroded and unusable when they arrived in the tropical theatres of war. To quote just one example from that period:

> Two aircraft were received at Cape Town after a two month voyage in slightly ventilated wooden packages . . . steel parts were heavily rusted, aluminium was corroded with some pitting. The wood was heavily discoloured . . . oozing growths were present on the wood.

There was initially a tendency to accept such incidents with resignation as 'tropical corrosion' but earlier successes of research, such as those of U. R. Evans of Cambridge University, into the mechanism and causes of corrosion fortunately offered a more constructive approach to these problems. Government scientists, with Evans and his colleagues, began a series of investigations which provided explanations of many of these war-time failures, and this new understanding allowed better practices to be developed which provided potential solutions to the problems of corrosion and deterioration. Improved codes of practice and specifications covering the selection of materials, protective finishes, methods of packaging, and design procedures were followed, and this led to a marked reduction in corrosion failures in Government equipment in the tropics and elsewhere.

Lest readers infer from these remarks that tropical conditions are intrinsically corrosive it should be noted that very low rates of corrosion are often recorded in tropical areas — the Delhi Pillar for instance is made of iron and has remained uncorroded for centuries despite being unprotected and fully exposed in a tropical area. Even in tropical jungles corrosion rates are often lower than in temperate industrial areas, but *when a metal is contaminated with certain active compounds and is then exposed to the hot humid conditions typical of wet tropical areas, then very rapid corrosion often ensues.*

It was known to those involved in transporting and storing UK equipment in World War 2 that corrosion increased with humidity and with temperature, so that some increase of corrosion rates was to be expected under warm damp conditions. It was however the haphazard occurrence of corrosion and especially the often devastating attack on sophisticated equipments, packaged to what was considered a high standard, which was the cause for most concern.

Rusting of steel components of the packaged aircraft quoted earlier was a typical example of this problem. These components had been protected with electrodeposited cadmium, a coating which had been selected after extensive trials because it had been shown to protect steel against corrosion for many years when fully exposed on a tropical surf beach — the worst conditions which could have been envisaged, and yet, when shielded from the weather in a protective package the cadmium coating had corroded through and the steel had been seriously rusted, all in a matter of two months. In this instance the effect was later shown to be due to acetic acid given off by the wood of the packing case. Timber at that time was in short supply and the relatively new process of kiln-drying wood was being introduced in place of the natural seasoning which had been the previous practice. Under the hot moist conditions of the kiln the acetyl groups which are bound within the cellulose of the wood had hydrolysed and released acetic acid which had little time to escape before the timber was incorporated into a package.

More will be said of the problems caused by wood and other sources of contaminants in the later chapters of this book, but before leaving the particular instance of the packaged aircraft, the 'oozing growths from the timbers' illustrates another aspect of the problems of corrosion. Corrosion of metals is often associated with deterioration of organic materials which may be part of the structure being protected, or of the packaging, because the warm damp conditions under which corrosion most readily occurs are well suited to both microbiological growth and to the hydrolysis of some of the less stable polymers. There is indeed often an interaction and sometimes an interdependence between the various processes of deterioration. The materials of the protective barriers or of the coatings may suffer biological degradation and cease to afford protection against corrosion; frequently the breakdown products from adjacent materials are corrosive and occasionally the products of corrosion accelerate reactions in surrounding materials. Because of this relationship between different forms of degradation, a summary account of the deterioration of materials other than metals has been included in this book, with emphasis on their roles in causing or accelerating corrosion of metals.

The discoveries of early investigations into the causes and cures of the corrosion of metals and the degradation of organic materials in storage and transit, allied with advances in the physical protection of items in transit and the development of new materials, laid the foundations of packaging and preservation as a technology in its own right. The industry sustained by this technology has matured over the last forty years to become one of the largest of modern industries in terms of capital investment, manpower employed, tonnage of goods handled and geographical extent.

The scientific studies of corrosion and the protection of metals have extended to match the breadth of problems which has arisen, although the extent of these activities is not at first evident since they are fragmented among the diverse component industries.

Understanding of corrosion and its prevention has progressed to the point where it is rare to find an incidence of corrosion that could not have been predicted and prevented if available advice had been followed; yet many corrosion failures still occur. One explanation often advanced is that it makes economic sense to reduce the money spent on corrosion prevention to the point where a small percentage of components will fail prematurely through corrosion. Although the force of this economic reasoning is difficult to resist, the increasing costs of raw materials and of fabrication, the greater emphasis on conservation, improvements in protective treatments, the frequently high consequential costs of failures in delays and breakdowns, and the greater discernment and sophistication of customers, all reduce the occasions when any corrosion can be tolerated before goods reach their final user. The vagaries of statistical predictions should not be forgotten; although probabilities of failure may be low, say 0.1%, these failures may be very embarrassing if clumping of these low probability events occurs in time or space and the items in one shipment or one week's production all corrode before they are in service.

More frequent reasons for corrosion failures are lack of knowledge and neglect. These twin reasons were identified as the major causes of unnecessary financial losses to corrosion by an official survey carried out by a committee of experts under the chairmanship of T. P. Hoar (1971) which reported on the costs of corrosion to the United Kingdom. This survey found that the direct annual cost of corrosion in the United Kingdom was at least £1300M, that a quarter of this sum could readily be saved, and that most industries continued to experience examples of serious corrosion. Although some instances were identified where new knowledge was needed, the cause of the continuing ravages of corrosion at such a high level was 'the general problem of poor communications and lack of awareness'.

A more recent Australian study using an econometric model estimated that A$2000M could be saved (at 1982 costs) within the Australian economy if known methods of corrosion protection were efficiently applied. This sum was equivalent to 1.5% of the Australian GNP and it did not include consequential damage.

The methods and materials available for protecting components have improved dramatically over the past thirty years with the developments of plastics and with the mechanisation of many processes which were previously either impracticable or were laborious and often unreliable since they had to be carried out by hand. Synthetic polymers now provide a range of versatile and robust materials for containers, barrier wraps, cushioning, supports and protective coatings. Grit blasting, metal spraying, paint spraying, electrophoretic coating, dip coating, solvent degreasing, vacuum evaporation and ion beam deposition are among the many processes now widely available in automatic or semi-automatic operation to aid the application of protectives. Over the same time, however, the tasks confronting those involved in protecting metals have multiplied with the advent of new technologies, the need for fast processing, the appearance of many new alloys and the tendency of designers to exploit to the full the load-bearing properties of metals so that some of the more intractable corrosion problems which involve the conjoint action of stress and corrosion, such as stress corrosion, corrosion fatigue and hydrogen embrittlement, are becoming more commonplace. The decreasing size of components in many applications, but especially in electronics, has made them ever more sensitive to corrosion. The critical nature of these reactions on the active elements of integrated circuits in particular is now an area of intensive study since it is a limiting factor in their long-term reliability.

The greater use of mechanical handling equipment, unit load containers and air-conditioned storage areas now allows the environments in which goods are stored to be maintained under favourable conditions for much of the time that they are in transit, and this has helped to reduce the hazards from corrosion. These facilities however are not always available at many smaller ports, and inspection and repackaging of the contents of containers in humid, wet and salty atmospheres is still commonplace; inspection at national boundaries by unskilled personnel may also leave containers unsealed and their contents at risk. Often the cost of providing good storage conditions is too high and protection must be attained by coatings and wraps. Knowledge and expertise is needed in selecting the appropriate method of corrosion prevention.

This book is intended to help bridge the gap between the corrosion expert and the concerned designer by providing a guide to the principles of the corrosion and protection of engineering metals which gives an account of the theory and then relates this to practice. A parallel intention is to provide practical guidance to those involved with the day to day protection of metals. References have been used sparingly in the text since the aim is to present a consensus rather than debate the issues from original papers; apologies are offered to any authors accidentally omitted.

Two principles have been followed in ordering the presentation. The first is to give a logical development from basic principles; the second is to keep the reader informed of the relevance of the principles to the practical problems of packaging. The next chapter contains a short description of the hazards, in which some of the major conclusions of subsequent chapters are anticipated, to give perspective to the theme of the subject. The chapters that then follow introduce the theoretical and experimental basis of corrosion, metal cleaning and protective treatments, with an emphasis on the environments encountered in storage and protection. The final section is devoted to the materials used in packaging. It has been assumed that the reader has some basic knowledge of chemistry and metallurgy, but an endeavour has been made to ensure that the book is intelligible and useful to those without this knowledge.

The books of Shreir (1976), Evans (1963, 1968 and 1976) and Parkins (1982) are recommended to readers seeking a full account of corrosion theory and practice. New information is published in a range of journals including *Corrosion Science, British Corrosion Journal* and *Transactions of the Institute of Metal Finishing*.

Standards are a major source of information referred to throughout the book. The British Standards are the major national standard. The Ministry of Defence, as a major customer, has its own series of guides and standards, which are widely available and often have a national status. The leading defence specifications are the Defence Standards (DEF STAN), but Defence Specifications (DEF), Defence Guides (DG), and DTD, CS and TS standards are also issued. International (ISO) Standards are also now more widely used, and in many instances are accepted by the British Standards Institution and some are published as British Standards.

Many other sources of information are available. The leading one in the field of corrosion, set up as a response to a proposal of The Hoar Report (1971), is the National Corrosion Service, National Physical Laboratory, Teddington, Middlesex, TW11 OLW. They offer a comprehensive advisory service and publish a valuable series of technical guides on aspects of corrosion.

Specifications and Standards are referred to frequently in the text and a comprehensive list of specifications, standards and guides related to the protection of metals in packaging is contained in the Annex at the end of the book.

REFERENCES

Evans, U. R. (1963) *An Introduction to Metallic Corrosion.* Edward Arnold.

Evans, U. R. (1968) *First Supplement.* St Martin's Press.

Evans, U. R. (1976) *Second Supplement.* St Martin's Press.

Hoar, T. P. (1971) *Report of The Committee on Corrosion and Protection.* HMSO.

Parkins, R. N. (1982) *Corrosion Processes.* Applied Science Publishers.

Shreir, L. L. (Ed.) (1976) *Corrosion.* Newnes-Butterworths.

2

Hazards

The ultimate result of good protective practice is nothing: no problems, no tarnishing, no corrosion rejects and no customer complaints, but equally: no added value and no appeals for more or better; possibly financial overseers may even be heard murmuring for less and cheaper. As these financiers will point out, money spent on corrosion protection above the essential minimum is money wasted. But insufficient attention paid to protection can result in complaints, problems, expensive reclamations, rejects, a valueless product and angry customers.

The aim of corrosion protection, as with other aspects of protection from deterioration in storage and transport, is to ensure sufficient protection but little more, and to provide it at minimum cost. Getting the balance right requires an overall knowledge of the hazards, and of the methods of combatting them through selection of materials, processing, protectives, storage conditions, and packaging methods and materials. Ignorance in these areas may not only lead to unnecessary expenditure but may give both high costs and high failure rates.

Ensuring that a product is capable of resisting corrosion throughout its service life is inevitably, and rightly, the responsibility of the Product Designer. Protection of raw materials and of components during manufacture is often left to the Production Engineer, and protection during delivery to the Package Designer. Although a degree of delegation is desirable, these three activities are all ultimately the responsibility of the Product Designer, and should be considered by him in conjunction with the Production Engineer and Package Designer from the earliest stage. It is advisable to start this task from basic principles and specifications, although due consideration should be given to previous experience with similar designs, and with the intended methods of storage and delivery. Successful corrosion protection is determined on the drawing board.

Although failures are costly and best avoided they can be turned into an asset if the correct lessons are learned and applied. Much of the theme of this book is an analysis of other people's failures, which, fortunately or unfortunately according to one's point of view, are likely to be long with us to be exploited in this way. A perceptive explanation of the continuing occurrence of examples of bad design or bad practice leading to corrosion failures is contained in the closing remarks of a valuable survey of *Classic Blunders in Corrosion Protection* by Oliver Siebert (1984) — 'We get so soon old, but so late smart'.

Experiments, experience and scientific deduction applied in a formal analysis of the requirements and environments likely to be encountered throughout the life of components is essential in identifying hazards and possible failure modes.

A first step in the selection of a protective scheme is to define the hazards to be encountered and their possible effects on the materials being used. This requires an analysis which determines the duration of exposure, the ambient environments, and the possible problems of compatibility likely to arise between the materials being used.

The hazards liable to be encountered may be broadly grouped as environmental and physical, and are summarised in Table 2.1.

Table 2.1 — Hazards in storage and transit.

Environmental (climatic)		Physical
Temperature extremes		Dropping
Temperature variation	Diurnal cycle	Bumping
Humidity		Vibration
Rain and other precipitation		Impact
Solar radiation		Puncture
Water immersion		Bending
Salt spray		Crushing
Dust, soot and solid debris		Lifting
Corrosive vapours		Dragging
Incompatible materials		
Wind		
Pressure variations		
Biodegradation		
Inspection, pilferage, and attack by rodents and insects		

CORROSION SUSCEPTIBILITY

Table 2.2 contains a summary of the conclusions of later chapters. It shows the range of metals encountered in general engineering (those most widely used are underlined) arranged in the first column on the basis of energy available for oxidation, and in the second column in an order which approximates to their resistance to general atmospheric corrosion. There is however a wider range of corrosion effects and some of these are indicated in the final column of Table 2.2

Susceptibility to some of the effects shown in Table 2.2 and their resulting failure modes, such as stress corrosion, whisker growth, and hydrogen embrittlement may be built in at the design stage; others such as corrosion by treatment residues may be caused by errors in processing, and still others, such as vapour corrosion or fretting, are more likely to arise from the packaging methods and materials used.

WATER AND HUMIDITY

Most corrosion requires the presence of water, either as liquid or vapour. Liquid water arises from precipitation as rain, snow, hail, mist, frost or dew, or, especially in marine atmospheres, splash and spray. Water vapour in the atmosphere supports corrosion with most aggressive contaminants when the relative humidity is above about 70%, but may continue at lower relative humidities with some contaminants; Evans (1972), for instance, showed that the prevention of metallic corrosion in the presence of sea salt requires a level of humidity at least as low as 35% and even lower if possible. In tropical environments relative humidities below these levels can also be hazardous since absolute water vapour pressures are high, and the 'breathing' of enclosures with the diurnal temperature cycle can introduce large quantities of water.

The greatest risk of water ingress occurs in transport by sea, dockside storage and storage in humid tropical conditions.

STORAGE AND TRANSIT REGIME

Table 2.3 is a summary of typical regimes of storage and transit which are liable to be met by goods during distribution.

The level of protection needs to be set to match the most hazardous series of conditions likely to be encountered. These are discussed in more detail in later sections but are summarised here to give perspective to the detailed account of corrosion and protective treatments contained in the following chapters.

High humidities and high temperatures are important factors in determining the occurrence and extent of both corrosion and biological processes of deterioration. The worst conditions occur in coastal tropical regions, and on the exposed decks of ships in the tropics, where temperatures are high during the day, falling at night, with large amounts of water vapour held in the atmosphere. A selection of temperatures and humidities is given in Table 2.4 for Bahrain, a typical tropical coastal area which presents a harsh storage environment.

The duration and intensity of sunshine increases temperatures within exposed packages, in buildings or immediately below ship's decks, well above that of the ambient air. Temperatures may reach 60°C in storage buildings, and 70°C in packages or containers exposed in the open, and still

Table 2.2 — Order of corrosion susceptibility of metals (most reactive at top).

Order of heat of oxidation available	Order of susceptibility to atmospheric corrosion (see Chapter 4) (engineering metals underlined)		Other serious corrosion hazards (see Chapters 5–9)
Magnesium	Iron	Protection necessary	VC CC F HE
Beryllium	Magnesium		VC CC SC
Titanium	Manganese		
Hafnium	Zinc		VC CC F Wh
Aluminium	Cadmium	Some corrosion	VC Wh
Zirconium	Rhenium		
Manganese	Lead		VC Wh
Vanadium	Vanadium	May need protection	
Chromium	Germanium		
Tantalum	Copper		SC OCC F HE
Niobium	Aluminium		SC F MC
Zinc	Nickel		SC
Gallium	Stainless steel (12–15% Cr)		CC SC HE
Indium	Tin	Some tarnish	Wh F Ph F
Germanium	Silver		F ST SWh HE
Tungsten	Cobalt		
Molybdenum	Tungsten		
Tin	Molybdenum		
Iron	Indium		VC
Cadmium	Gallium		
Cobalt	Stainless steel (above 18% Cr)		SC
Nickel	Chromium		
Rhenium	Beryllium		
Lead	Hafnium		
Copper	Zirconium		
Osmium	Osmium		
Silver	Titanium	Little effect	SC HE
Palladium	Palladium		
Rhodium	Platinum		
Platinum	Iridium		
Iridium	Gold		
Gold	Tantalum		
	Niobium		
	Rhodium		

Key to 'Other hazards':

VC Vapour corrosion
CC Contact corrosion
SC Stress corrosion
F Fretting corrosion
HE Hydrogen embrittlement
Wh Whisker growth
OCC Organic contact corrosion
MC Corrosion by mercury
Ph Phase change
ST Sulphide tarnishing
SWh Sulphide whisker growth

higher levels in especially adverse conditions; 90°C, for instance has been recorded in an aircraft cockpit exposed in full tropical sunshine.

The type of transport to be used is a major factor in defining the hazard level. The most widely used inland systems in order of increasing hazard are:

— full containers;
— mixed loads in own transport;
— public road services;
— mixed goods by rail;
— passenger rail and parcel.

Table 2.3 — Typical time and environments in storage and transport (basis of hazard analysis).

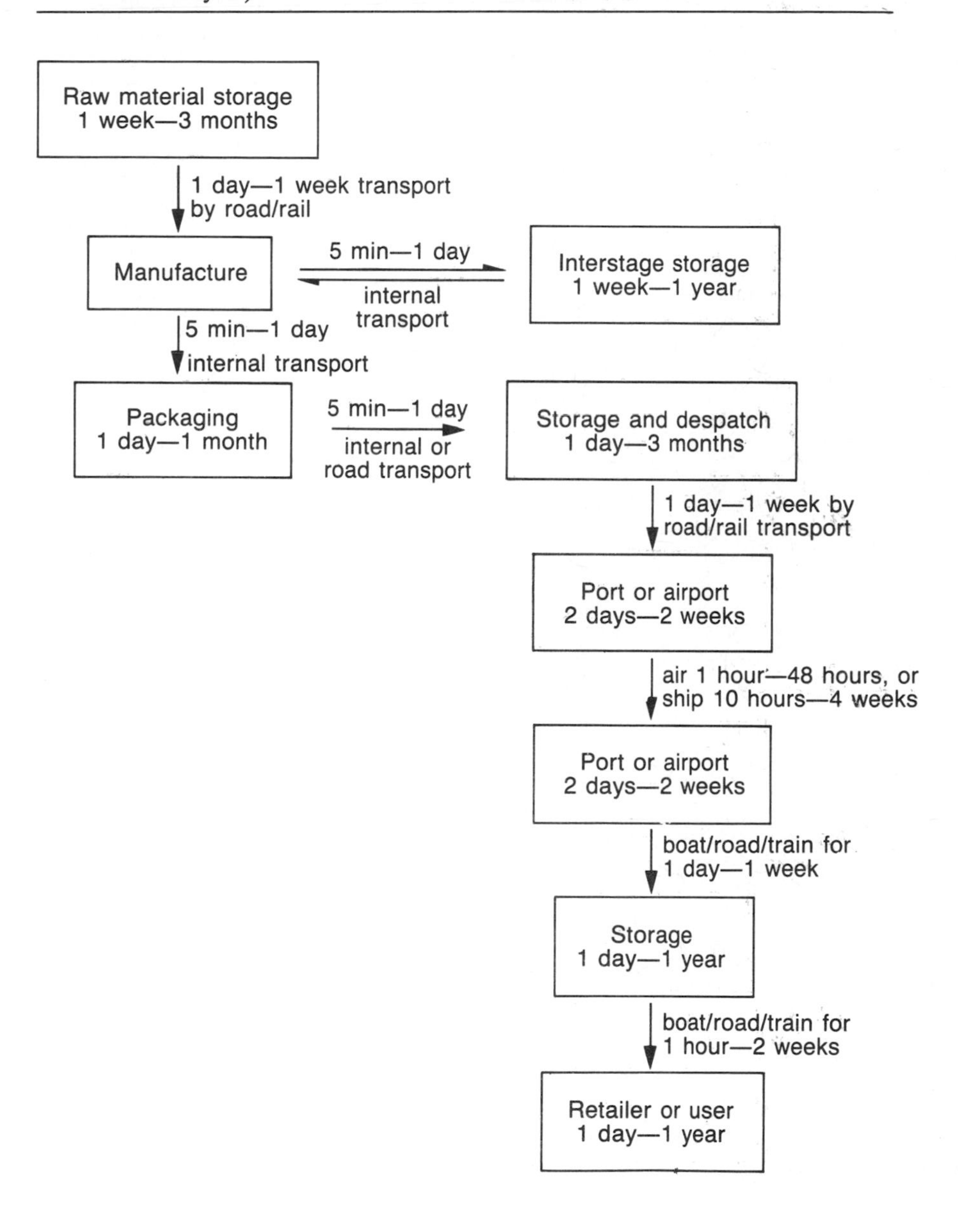

Table 2.4 — Temperatures, humidity and precipitation in Bahrain, Persian Gulf (26°12′N, 50°30′E). 1928—1943.

Period	Temperature (°C)						RH%		Precipitation
	Average				Absolute		Average of observations at		Average no. of days with 2.5 mm or more
	daily		monthly						
	Max.	Min.	Max.	Min.	Max.	Min.	07.30	15.30	
January	20	14	26	9	29	5	85	71	1
February	21	15	29	11	34	7	83	70	2
March	24	17	32	12	35	11	80	70	1
April	29	21	37	16	41	13	75	66	1
May	33	26	39	21	42	19	71	63	<0.1
June	36	28	40	18	43	21	69	64	0
July	37	29	42	26	44	24	69	67	0
August	38	29	42	26	45	24	74	65	0
September	36	27	40	23	44	22	75	64	0
October	32	24	36	21	39	19	80	56	0
November	28	21	33	17	36	14	80	70	1
December	19	16	28	12	31	9	85	77	2
Year	85	72	43[a]	8[b]	45	5	77	68	8
No. of years	16	16	16	16	16	16	16	9	16

[a] Average of highest each year.
[b] Average of lowest each year.

On ships, conditions range from mild to severe; from well stowed and sealed containers, through mixed loads of varying hazard to deck cargoes. Loading and unloading, temporary storage in the vicinity of ports and final distribution are all high risk conditions.

It will be seen from later discussion that the corrosion hazard is often built into a design through contaminants left on the surface or derived from packaging materials. Subsequent storage in damp, or worse hot damp, conditions stimulates corrosion. Hazards are also presented by contaminants from industrial atmospheres, boiler fumes, treatment solutions, marine mist, or from other goods contained within the same enclosure. Rodents, termites and mould growth damage packaging and they are also a source of corrosive detritus; their proliferation is favoured by warm damp conditions.

Mechanical damage from dropping, bouncing, stacking, vibration and mishandling may damage outer containers, wraps, barriers and supports, so that water vapour and contaminants may enter. Pressure changes in air-flights and wide variations in temperature such as are experienced in the daily cycle of temperature and humidity in many hot climates and also within the holds of ships passing from cold to warm climates, induce 'breathing' of packages and consequent condensation, especially on items of high heat capacity. Sea mist and storms at sea, and dust storms on land, may introduce contaminants, especially if seals have been broken. On this last point pilferage and customs inspection as well as rough handling frequently result in damage to seals.

CORROSIVE CONTAMINANTS WITHIN ENCLOSURES

Industrial and domestic effluents rich in sulphur dioxide and soot, and sea salt are the contaminants responsible for most of the corrosion of metals exposed in open environments. However in enclosures such as packages, containers, ships' holds, storage areas and workshops, the access of external air with its particulate and gaseous contaminants is restricted and has to flow through narrow inlets. Under these conditions little particulate matter enters, and gaseous contaminants in the atmosphere are reduced by absorption on the surfaces of the narrow entry paths through the packaging so that few corrosive impurities penetrate to sensitive surfaces within enclosures from the outside atmosphere — the sulphur dioxide levels within buildings for instance are usually well below that of the external atmosphere and if the doors and windows of a room are even moderately well sealed the concentration of sulphur dioxide in the air soon falls to a negligible level. It is however a matter of experience that corrosion does occur in enclosures, often at a greater rate than in the external atmosphere. In the great majority of such instances the contaminants responsible for the attack are either already on the metal surfaces or are derived from other materials within the enclosure. Contaminants likely to be encountered are numerous but may be considered under the following headings:

— vapours of organic acids derived from materials within the enclosure (Chapter 6);
— contact with materials containing corrosive ions which are either already present or are derived from paints, temporary protectives, wrappings, packaging materials, or from parts of the assembly itself (Chapter 7);
— retained residues from manufacture and treatment solutions, e.g. soldering flux, handling (sweat), and organic crack detection and degreasing fluids (Chapter 8);
— activities of animals, moulds or bacteria (Chapter 9);
— deposits from splash, condensation or leaks which have leached out contaminants.

Both the humidity and contaminants within an enclosure are thus often different from those in the surrounding atmosphere. Any policy for whole life protection requires a total strategy which includes an adequate control of all the essential ingredients of the corrosion process throughout the life of the item — from the time when the metal is first extracted from its ore to the point of delivery to the final user. During transit and storage the highest standard of protection is obtained by maintaining the humidity below the critical corrosion threshold or by preventing access of oxygen by the use of an impervious container sealed in an atmosphere of dry nitrogen or under vacuum. When neither of these methods is practicable because of costs or the need for frequent access for inspection or maintenance, then a somewhat lower standard of protection is obtained through the use of impervious wraps and a sealed container; a lower although often adequate level of protection is provided by temporary protectives and wraps within an unsealed container. However for economy or for rapid access, often only an unsealed external container without wraps is used, and occasionally, as with heavy machinery or motor cars, the only protection provided is from temporary protectives.

CORROSION HAZARD TESTING

A range of tests is available for simulating some of the hazards which may occur in service. A typical series will be found for instance in British Standard 2011 'Basic environmental testing procedures'. Three tests have been found especially useful by the writer for evaluating corrosion hazards and are referred to frequently throughout the text. These are:

The salt droplet test (BS 1391). This gives a simple and rapid classification of the effectiveness of protective coatings.

Damp heat testing, typically at 30°C and 100% RH, for revealing any incompatibility between materials and the presence of corrosive residues. The effect of mould growth may be evaluated at the same time by the introduction of a suitable selection of spores at the start of the test.

Hot damp cycling test. That most frequently used by the writer has been Ministry of Defence, Intensified Standard Alternating Trial, condition A (ISAT(A)) of Defence Standard 07–55 (Part 2). The weekly cycle of ISAT(A) is:

3 days, 46°C, RH above 95%.
1 day cooling to ambient.
1 day, 46°C, RH above 95%.
1 day, 60°C, RH 60% ±2%.
1 day cooling to ambient.

This test is useful for packaged equipment as it stimulates the breathing effect which may be so damaging in humid atmospheres. The test reveals compatibility hazards between materials of the package and may give an indication of latent problems from ageing and hydrolysis. For temperate environments these conditions may be overexacting and the temperatures should be reduced by about 16°C.

External and covered exposures, and test pieces within packages transported over the known route give useful supplementary information, but simple tests with an informed and intelligent assessment of all the effects observed are usually sufficient.

REFERENCES

Siebert, 0. (1984) Coburn, S. K. (Ed.) *Corrosion Source Book*. Am. Soc. for Metals & NACE.
Evans, U. R. (1972) *Br. Corros. J.* **7**, 10.

FURTHER READING

BS 4672 *Guide to Hazards in the Transport and Storage of Packages.*
BS 5073 *Stowage of Goods in Freight Containers.*
BS 1133 *Packaging Code.*
Defence Guide-11 *Design Requirements for Service Packaging.*
DEF STAN 07-55 (Pt 2) *Climatic Environmental Conditions Affecting the Design of Materiel for use by NATO Forces in a Ground Role.*

3

Why and How Metals Corrode

HEATS OF FORMATION

Metals, other than the precious metals, seldom occur naturally in the earth's crust in an elemental metallic form but have to be extracted from compounds (ores) using heat, chemical, electrical or mechanical energy. Once extracted, metals start the return to their 'native' state by reacting with the oxygen in the air, releasing heat as they do so. The energy that is available for this reaction can be expressed in terms of the heat evolved when the metal reacts with oxygen, or often more conveniently, as the electrical potential between a metal and an aqueous solution of a standard concentration of the ions of the metal.

Iron, aluminium, zinc and magnesium are the widely encountered constructional metals and although they corrode readily outdoors when unprotected, when painted they will endure for long periods, even in corrosive environments, and in dry air they will last indefinitely without protection. But if ignited, a ribbon of magnesium metal burns vigorously to give a white powder of magnesium oxide. Finely divided iron, zinc and aluminium also burn although not so readily as magnesium.

THE OXIDE FILM

The vigorous reaction of these metals when ignited is in contrast to their apparent stability in air, but this stability is in a sense illusory since any freshly formed surface of these metals exposed to the air reacts immediately with the oxygen to form an invisible layer of oxide. This thin but complete covering of oxide then acts as a barrier between the metal and the air. The extent of any further reaction between the metal and oxygen is then determined by the properties of the oxide layer, which in many environments is stable and impervious to oxygen so that the metal is protected. When the metal is ignited it melts and then vapourises so that the oxide no longer forms a barrier and cannot prevent the metal from mixing and reacting with the atmosphere.

ELECTROCHEMICAL DISSOLUTION

Essentially the same conversion of metals to their oxides, or their hydrated forms, the hydroxides, can occur when the metal is immersed in an electrolyte at constant temperature. The energy is then released as electricity instead of heat. Fig. 3.1 shows the reactions involved when metallic zinc corrodes in an electrolyte.

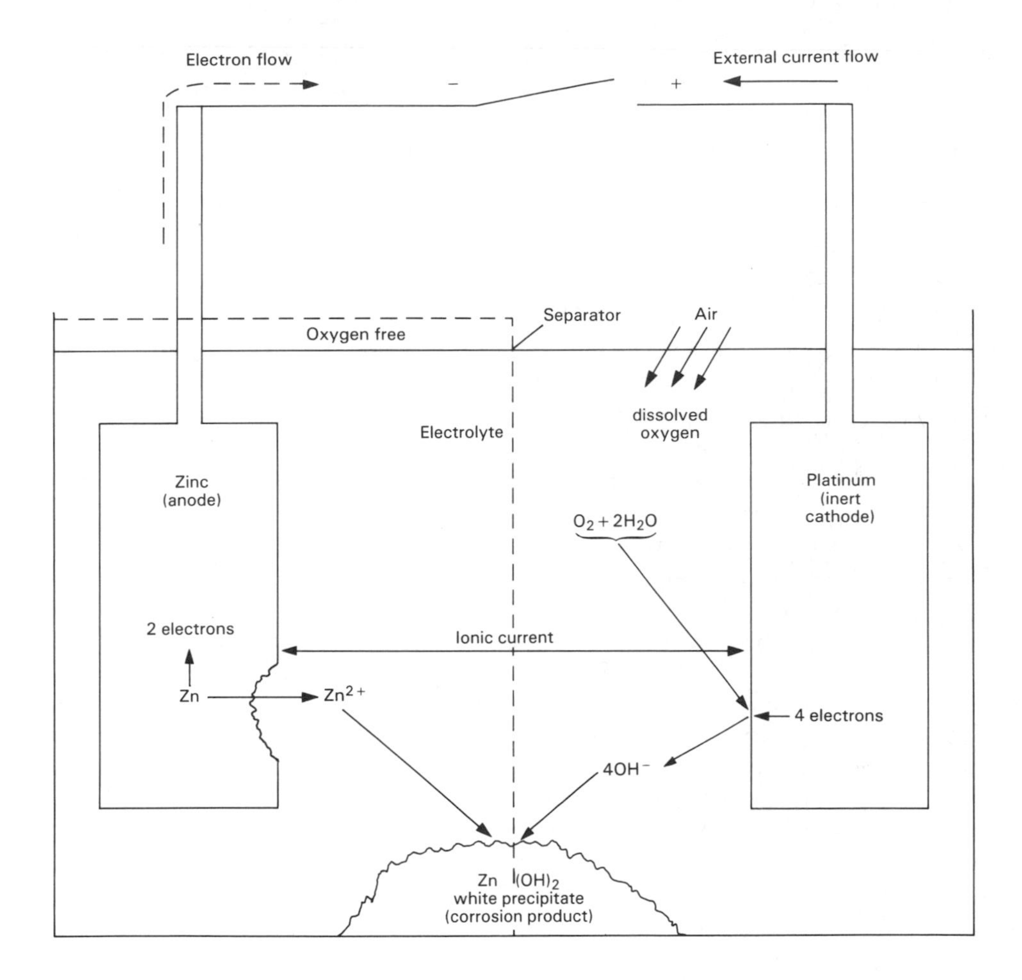

Fig. 3.1 — Electrochemical dissolution of zinc in aerated electrolyte.

In the system represented in Fig. 3.1 the two major reactions have been separated; the principle is essentially that exploited in commercial electrochemical cells. At the zinc electrode, the metal passes into solution as positively charged ions, leaving a negative charge (electrons) on the metal.

i.e. $Zn \rightarrow Zn^{2+} + 2e$

The positive metal ions are known as **cations**; the zinc electrode is the **anode.** If the external circuit between the two immersed electrodes is not completed then the release of zinc ions from the anode soon ceases, as the negative charge builds up on the metal and creates a potential barrier between the metal and the solution. This potential barrier, which exists as a double layer of hydrated positive ions on one side and electrons within the metal lattice on the other, hinders the further escape of ions from the metal and attracts back zinc ions from the solution. When the metal is immersed in the electrolyte a kinetic equilibrium is rapidly established with the anode exhibiting a constant negative potential in equilibrium with a constant concentration of ions in solution.

The inert platinum electrode in Fig. 3.1 is immersed in the electrolyte which contains dissolved oxygen. This oxygen can react with water at the surface of the electrode, extracting electrons from the platinum surface as it does so, and giving soluble hydroxyl ions in solutions:

i.e. $O_2 + 2H_2O + 4e \rightarrow 4OH^-$

The negatively charged ions are known as **anions;** the electrode at which they form is the **cathode.** If the external circuit is not completed an equilibrium condition is soon attained with a positive charge on the electrode.

For the cell shown in Fig. 3.1, if the electrolyte around the zinc electrode contains a molar concentration of zinc ions, and that around the platinum electrode contains unit activities of hydroxyl ion in equilibrium with oxygen at a pressure of one atmosphere,

i.e. $Zn/1N\ Zn^{2+}/O_2(1\ atm)/OH^-$ (unit activity)

then at 25°C the open circuit potential difference across the cell is 1.16 volts.

When the external circuit is connected between the two electrodes in the cell shown in Fig. 3.1, electrons flow in the external circuit from the zinc to the platinum and continuing reaction is possible, with zinc going into solution at the anode, oxygen being reduced to hydroxyl ions at the cathode, and a current flowing in the external circuit. The hydroxyl ion concentration near the cathode makes the solution more alkaline (increases the pH) in that region. As the concentrations of zinc and hydroxyl ions build up in solution the solubility of zinc hydroxide is soon exceeded and a precipitate forms,

i.e. $Zn^{2+} + 2OH^- \rightarrow ZnO.H_2O$ (precipitate)

Thus overall, zinc metal has reacted with oxygen from the air to give a hydrated zinc oxide, a result which is similar to that obtained from burning zinc in air, but the surplus energy has been released as electrical energy rather than heat.

When a connection is made in the cell shown in Fig. 3.1 the oxygen content of the solution near the cathode is reduced and the cell voltage and the external current fall; the electrode is said to be 'polarised'. The current is controlled by the rate at which oxygen dissolves and diffuses through the water to the cathode. Polarization is overcome in commercial cells by surrounding the cathode, which is usually a carbon rod, with a conducting paste of manganese dioxide and carbon powders. The reduction of manganese dioxide supports a steady current — the manganese dioxide is sometimes said to be functioning as a 'depolarising agent'.

ELECTRODE REACTIONS

The electrochemical conversion of zinc to its oxide takes place through two electrode reactions: the oxidation of zinc metal to zinc ions at the anode, and the reducton of dissolved oxygen to hydroxyl ions at the cathode (readers should note the convention if they are not already aware of it that a loss of electrons, or an increase of the electropositive nature of a substance, is known as reduction while an increase of electrons or an increase in the electronegative character of a substance is known as oxidation, although oxygen is not necessarily involved). This system of two electrodes, and the accompanying reactions is termed 'a cell'.

The potential across the electrodes of a cell when no current is flowing in the external circuit, is the total effect of these two reactions, each of which is known as a half-cell. It is not possible to measure the potential of a single half-cell, as any measuring system involves solid/liquid interfaces which introduce additional potentials. However it is possible to measure the potentials of a series of half-cells each in conjunction with one particular half-cell, so that comparative values are obtained. This procedure has been standardised and the hydrogen half-cell has been agreed as the electrode against which other half-cell potentials are quoted. This standard 'zero potential' is defined as the potential that exists under reversible conditions between hydrogen gas at one atmosphere and hydrogen ions at unit activity.

$$\text{i.e. } H_2 \text{ (1 atmosphere)}/H^+ \text{(unit activity)} = 0 \text{ volts}$$

The standard hydrogen electrode is arbitrarily defined as at zero potential for all temperatures. In practice it can consist of a platinum black electrode immersed in an acid solution which contains hydrogen ions at unit activity and with hydrogen gas bubbling through; since platinum black is an efficient catalyst for this reaction the potential remains close to equilibrium conditions. It is however usually convenient to use a more easily prepared half-cell as a sub-standard, and an electrode of silver, coated with silver chloride in a solution of chloride ions at unit activity, is widely used in this role (at 25°C the potential of this Ag/AgCl (solid)/1N Cl^- electrode against a standard hydrogen electrode is +0.2224 volts).

CONCENTRATION POTENTIALS

The electrochemical reactions considered so far are those which arise at solid/liquid interfaces, but any boundary which separates differing concentrations of ions has a potential step associated with it. Looking back at Fig. 3.1 for instance: it was assumed that the concentration of dissolved oxygen was high around the cathode and low around the anode. This can be arranged physically by placing a separator between the two areas and allowing air to bubble around the cathode and nitrogen around the anode; a concentration potential then exists across the separator.

Similar concentration cells arise if a silver/silver chloride electrode is used for measuring half-cell potentials in solutions containing low concentrations of chloride, since the change of concentration of chloride ions across the interface between the standard concentration of chloride ions of the half-cell and the solution around the electrode of interest gives an additional potential to that being measured. For monovalent ions these concentration potentials are 59 mV for each tenfold change of concentration, and although only a second-order effect in overall cell reactions they complicate accurate measurements and are often important in corrosion reactions.

HYDROGEN ION CONCENTRATION: pH

As the hydrogen ion concentration in aqueous solution can vary over a very wide range, being about 1 g per litre in 1N acid solution and 10^{-14} g per litre in a 1N solution of alkali, it is conveniently expressed as a logarithmic function; for less obvious reasons the chosen function, the pH, is the negative logarithm to the base 10 of the hydrogen ion concentration:

$$\text{i.e pH} = -\log[\text{concn H}^+], \text{ or conc H}^+ = 10^{-\text{pH}}$$

Thus alkaline solutions have high pH values while acid solutions have low pHs. The concentrations of hydrogen ions and hydroxyl ions are in equilibrium in aqueous solution, and are equal at 10^{-7} gl^{-1} in neutral solutions. The relationship between pH, and pOH, the corresponding function for the hydroxyl ion, is:

$$\text{pH} + \text{pOH} = 14$$

It should be noted that when the concentration of hydrogen ions exceeds 1 g per litre the pH is negative.

(Strictly, activities rather than concentrations are involved in electrochemical reactions, but for an account of the significance of the difference the reader is referred to electrochemical texts; an understanding is not needed for the level of treatment in this text.)

ELECTROLYSIS OF WATER AND OVERPOTENTIAL

On some metals the potential at which hydrogen is released at measurable rates is substantially displaced from the reversible potential; the phenomenon is known as overpotential. The magnitude of the overpotential of hydrogen release from water sometimes allows other electrode reactions to occur in solution even though energetically hydrogen release may be the preferred reaction.

The cause of the high overpotential of hydrogen lies in the complexity of the reactions which lead to hydrogen release. One or two of the reaction steps proceed more efficiently on some surfaces than on others; the noble metals, particularly finely divided platinum black, give favourable conditions for these reactions and have low overpotentials. The critical steps of hydrogen discharge are: neutralisation of the hydrogen ion at the metal surface to give neutral atoms, combination of pairs of atoms to give hydrogen molecules, and the agglomeration of the molecules into bubbles.

$$\text{i.e. } H^+ + e \rightarrow H \text{ (atom)} \quad (1)$$

$$2H \rightarrow H_2 \text{ (molecules)} \quad (2)$$

$$H_2 \text{ (molecules)} \rightarrow \text{gas bubbles} \quad (3)$$

When one or more of these steps is inhibited there is no significant release of hydrogen until the potential is taken well beyond the reversible potential; this increment is the overpotential. On pure iron and on zinc, overpotential is high at about 0.25V, but is lower on impure samples of these metals. Hydrogen overpotential is low on platinum, palladium, silver, gold, copper and carbon.

Hydrogen overpotential is often important in corrosion reactions. The shelf life of zinc batteries for example is dependent on overpotential, since fortunately, release of hydrogen is inefficient on zinc surfaces with their high overpotential. The overpotential can be even further increased by treating the zinc surface with mercury, and zinc anode surfaces in commercial cells are usually treated with mercury to obtain this advantage. Hydrogen release can be further decreased by lowering the hydrogen ion concentration in solution, as has been achieved with the use of potassium hydroxide based electrolyte in the zinc alkaline cells which are rapidly displacing the older acid chloride cells. This line of logic has been further exploited by moving away from aqueous electrolytes to polar organic solvents free of hydrogen ions in which lithium, which displaces hydrogen too readily from water to be used in an aqueous electrolyte, can be used as the anode. With appropriate depolarisers, and additions to increase conductivity, these lithium cells can give up to 4 volts per cell and have a high capacity and long storage life.

The problems of hydrogen embrittlement which are discussed in Chapter 5 are often exacerbated through hydrogen overpotential.

Other reactions sometimes exhibit overpotentials, but of those normally encountered only that of oxygen release is comparable in size to hydrogen overpotential.

ELECTRODE POTENTIALS

The reversible electrode potentials of metals are important in electrochemistry and in corrosion studies, and where they cannot be measured directly they are calculated from thermochemical reaction data. The normal reversible electrode potentials for the more commonly encountered metals are given in the 3rd column in Table 3.1.

The metals at the top of Table 3.1 have most energy available for reaction and they might therefore be expected to corrode most rapidly.

Table 3.1 — Free energy and electrochemical potentials of metals

Metal	Chemical reaction	Free energy of formation (kcal per g mole of oxygen)	Electrochemical reaction	Electrode potential e_0 (25°C) V	Passive/non-reactive pH range in aqueous solution
Lithium			Li/Li^+	− 3.0	Nil
Potassium			K/K^+	− 2.9	Nil
Calcium	$2Ca + O_2 \rightarrow 2CaO$	290	Ca/Ca^{2+}	− 2.87	Beyond 14.5
Magnesium	$2Mg + O_2 \rightarrow 2MgO$	275	Mg/Mg^{2+}	− 2.38	Beyond 11.4
Beryllium	$2Be + O_2 \rightarrow 2BeO$	275	Be/Be^{2+}	− 1.80	3.9–10.6
Aluminium	$\frac{4}{3} Al + O_2 \rightarrow \frac{2}{3} Al_2O_3$	247	Al/Al^{3+}	− 1.70	3.9–8.6
Titanium			Ti/Ti^{2+}	− 1.63	2.0–12.0
Vanadium	$2V + O_2 \rightarrow 2VO$	214	V/V^{2+}	− 1.2	Slight
Manganese	$2Mn + O_2 \rightarrow 2MnO$	180	Mn/Mn^{2+}	− 1.10	10.5–13.0
Zinc	$2Zn + O_2 \rightarrow 2ZnO$	149	Zn/Zn^{2+}	− 0.76	8.4–10.5
Chromium	$\frac{4}{3} Cr + O_2 \rightarrow \frac{2}{3} Cr_2O_3$	179	Cr/Cr^{3+}	− 0.7	4.4–12.5
Gallium			Ga/Ga^{3+}	− 0.52	3.0–11.3
Iron	$2Fe + O_2 \rightarrow 2FeO$	114	Fe/Fe^{2+}	− 0.44	9.5–12.2
Cadmium	$2Cd + O_2 \rightarrow 2CdO$	112	Cd/Cd^{2+}	− 0.40	9.8–13.5
Cobalt	$2Co + O_2 \rightarrow 2CoO$	101	Co/Co^{2+}	− 0.28	9.2–13.0
Nickel	$2Ni + O_2 \rightarrow 2NiO$	100	Ni/Ni^{2+}	− 0.23	9.0–12.1
Tin	$2Sn + O_2 \rightarrow 2SnO$	118	Sn/Sn^{2+}	− 0.14	− 0.5–12.5
Lead	$2Pb + O_2 \rightarrow 2PbO$	89	Pb/Pb^{2+}	− 0.12	8.5–11.0
Copper	$4Cu + O_2 \rightarrow 4Cu_2O$	70	Cu/Cu^+	+ 0.52	6.9–12.7
Silver	$4Ag + O_2 \rightarrow 2Ag_2O$	5	Ag/Ag^+	+ 0.80	Low reactivity
Rhodium	$4Rh + O_2 \rightarrow 2Rh_2O$	38	Rh/Rh^{3+}	+ 0.8	Passive/immune
Palladium	$2Pd + O_2 \rightarrow 2PdO$	40	Pd/Pd^{2+}	+ 0.83	Immune
Platinum			Pt/Pt^{2+}	+ 1.2	Immune
Gold	$\frac{4}{3} Au + O_2 \rightarrow \frac{2}{3} Au_2O_3$	− 1.0	Au/Au^+	+ 1.4	Immune

Those at the foot have little or no energy available for reaction and show little tendency to corrode; as they remain uncorrupted they are known as the Noble Metals. Although a cursory glance at the order of metals in Table 3.1 will confirm the general truth of this order of reactivity, a more detailed study shows numerous anomalies. Thus, for example, titanium is a metal noted for its corrosion resistance and is widely used in chemical plant for handling corrosive chemicals, and yet it has a high electrode potential, well above zinc and iron, two metals which readily corrode in many everyday environments. Aluminium and chromium are likewise higher in the table than their widespread use as corrosion-resistant metals would appear to suggest. To understand and explain these apparent anomalies it is necessary to consider the reaction mechanism of corrosion.

PASSIVATION

Elements of the corrosion reaction have already been indicated in describing the dissolution of zinc in Fig. 3.1. Corrosion occurs through several related simple reactions, some of which are in sequence and others in parallel. Like all chemical reactions the overall rate is mainly controlled by the slowest step within the fastest overall set of reaction steps.

The first stage of corrosion occurs as soon as a new metal surface forms and is exposed to the atmosphere, when an oxide film develops within a fraction of a second. The initial film is only a few atoms thick and at normal ambient temperature and in dry conditions it separates the metal from the air. The reaction between metal and oxygen may then be restricted as the oxide thickens at ambient temperatures.The rate of further reaction then depends on the properties of this oxide barrier which, while it remains may or may not protect the metal from the atmosphere.

The conditions under which the oxide layer becomes unstable are therefore those under which corrosion becomes probable. In pure water under neutral conditions only the oxides of very reactive metals such as sodium, potassium, calcium and magnesium are soluble; these oxides dissolve to give alkaline solutions:

$$\text{i.e. } K_2O + H_2O \rightarrow 2KOH$$

If soluble ions are dissolved in the water other reactive metal oxides can dissolve as soluble salts, especially in acid conditions,

$$\text{i.e. } FeO + 2H^+ (\text{acid solution}) + SO_4^{2-} \rightarrow FeSO_4 + H_2O$$

The ease with which the oxide is dissolved to give metal cations depends on the pH of the solution and the anions present; chlorides are especially effective in destabilising the oxide film on most metals.

The dissolution of metal oxides is favoured by acid conditions and becomes less probable as the acidity decreases; the oxide is seen to be reacting as an alkali. However some metal oxides, known as amphoteric oxides, may also react with alkalis to give soluble complex cations, so that these oxides will react as either acids or alkalis. Aluminium oxide is a typical amphoteric oxide.

$$\text{i.e. } Al_2O_3 + 2OH^- \rightarrow 2(AlO_2)^- + H_2O \text{ (oxide reacts as acid)}$$

$$\text{and } Al_2O_3 + 6H^+ \rightarrow 2Al^{3+} + 3H_2O \text{ (oxide reacts as base)}$$

Zinc and iron are two other common metals with amphoteric oxides. The readiness with which metal oxides can react as either an acid or an alkali to give soluble products is reflected in the range of pH over which the oxide will dissolve. Thus the thin film of ferrous oxide on steel will become unstable in electrolytes under acid conditions and in solutions of pH values up to 9.5, reacting as an alkali, whereas zinc oxide being slightly less basic will dissolve in solutions up to a pH of 8.4. Aluminium oxide is a weak base dissolving only in acid solutions with pH values up to 3.4. These three oxides will also react as acids to form soluble complex cations at high pHs: ferrous oxide dissolving with increasing readiness as the pH rises above 12.2, with zinc oxide dissolving when the pH rises above 10.5 and aluminium oxide above a pH of 8.6.

i.e.

$$FeO + OH^- \rightarrow HFeO_2^- \text{ Dihypoferrite (above pH 12.2)}$$

$$ZnO + OH^- \rightarrow HZnO_2^- \text{ Bizincate (above pH 10.5)}$$

$$Al_2O_3 + 2OH^- \rightarrow 2AlO_2^- + H_2O \text{ Aluminate (above pH 8.6)}$$

The dihypoferrite and bizincate ions are converted to hypoferrite (FeO_2^{2-}) and zincate (ZnO_2^{2-}) respectively as the pH increases further.

Looking at the solubilities of the oxides from an opposite view, the range of pHs over which they are stable and will remain to protect the metal from attack (an effect known as **passivation**), we have ranges of:

iron oxide stable, and therefore passivates, pH 9.5–12.2
zinc oxide stable, and passivates, pH 8.4–10.5, and
aluminium oxide stable, and passivates, pH 3.9–8.6.

The wider range of stability of aluminium oxide at near neutral pHs, typical of the weakly acid solutions of atmospheric pollutants, is one reason why aluminium resists atmospheric corrosion more readily than iron or zinc, although it is higher in the electrochemical series and therefore has more energy potentially available for reaction. Titanium is close to aluminium in the electrochemical series but its even wider passive range, from pH 2.0 to 12.0, accounts for the apparent inertness of this metal. In column 3 of Table 3.1, the ranges of pH over which the metals are passive are listed and these values help in assessing the likelihood of a metal reacting in a solution of given pH.

Other factors however modify the stability of the oxide and either increase or decrease the tendency for the oxide to dissolve; among these are certain reactive anions, particularly chlorides, and oxidising, reducing or complexing agents. Oxidising agents increase the stability of oxide films while reducing agents hinder their repair and tend to destabilise them. Dissolved oxygen from the air is usually present in any film of solution which forms on metals under atmospheric conditions and this helps stabilise the protective oxide coating. Complexing agents will remove metal ions and metal oxide cations from the solution locking them up as unreactive complexes; this encourages the dissolution of further metal to restore the ionic balance. The physical nature of the metal surface also affects passivation; smooth clean surfaces passivate more readily than rough, contaminated or stressed surfaces.

INHIBITORS

Some soluble additives reduce the risk of corrosion by modifying the metal surface, either stabilising or reinforcing the oxide film, or increasing the polarisation of the cathodic reaction. The former mechanism is known as anodic inhibition, the latter is cathodic inhibition.

Chromates are efficient anodic inhibitors for many metals particularly for steel, copper, zinc, cadmium, magnesium and aluminium. Some organic acids and bases are useful inhibitors; benzotriazole, an efficient anodic inhibitor on copper alloys, whose effects taken from a study by Donovan and Heron (1971) are shown in Fig. 3.2, is a typical example. The much lower current in the presence of benzotriazole as the copper electrode is anodically polarised (positive charge increased) shows that the anodic process is inhibited; there is some effect on the cathodic reaction, but this is small in comparison to the anodic effect.

Salts of fatty acids, which develop from reaction between metal oxide pigments and organic acids formed during the curing of oleoresinous paints have been shown by Mayne (1970), and Appleby & Mayne (1967), to give good inhibition; their action involves both anodic and cathodic mechanisms but other effects such as the incorporation of corrosive ions into insoluble reaction products are probably also involved.

POTENTIAL–pH DIAGRAMS

The reactions of metals and oxide films in contact with solutions over a wide range of pH and electrochemical potentials have been conveniently summarized in potential–pH diagrams, where the limits of the zones of passivation, immunity and corrosion are depicted in terms of pH and potential. These potential–pH diagrams have been developed by Marcel Pourbaix with his collaborators (1966), who have published detailed diagrams covering most metals in a wide range of reactions.

The potential–pH diagrams are elegant and informative summaries of

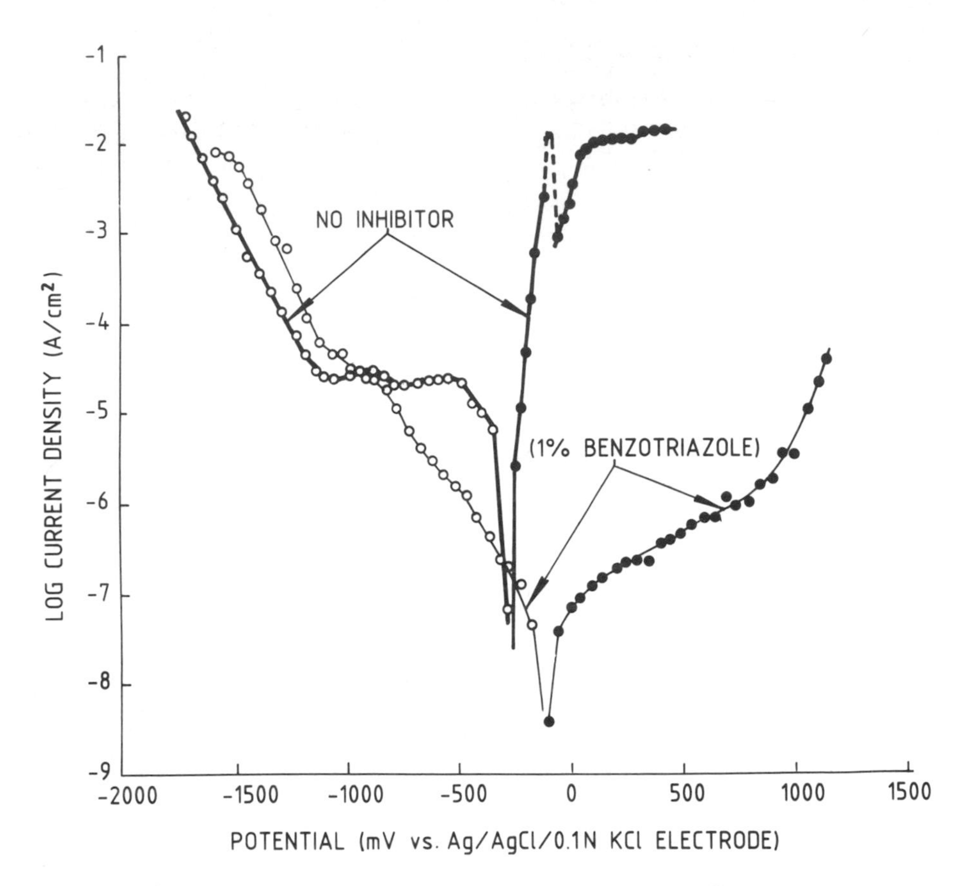

Fig. 3.2 — Polarization curves for copper in aerated 3% NaCl. (© Controller, Her Majesty's Stationery Office 1986.)

much useful and relevant information on the corrosion susceptibilities of metals but they are silent on rates of reaction and on many of the effects determined by the physical and electrical properties of the metal oxide films. Since it is usually the rates of reaction which are of final interest in corrosion studies, the role of these diagrams should not be exaggerated but neither should their utility be ignored. For the present however it should be recognised how well they account for the apparent anomalies in reactivity which are evident when metals are listed on the basis of electrochemical potentials. In particular the wide passive ranges of beryllium, aluminium, titanium, chromium and tin account for the corrosion resistance of these metals. The passivating effect is often carried over to alloys containing a proportion of these metals, so that alloys of iron and chromium for instance give 'stainless' steels if more than 12% of chromium is present.

CORROSION IN SOLUTION

The electrolytic mechanism for the corrosion of iron in a static neutral solution is illustrated in Fig. 3.3. The similarities with the zinc cell (Fig. 3.1)

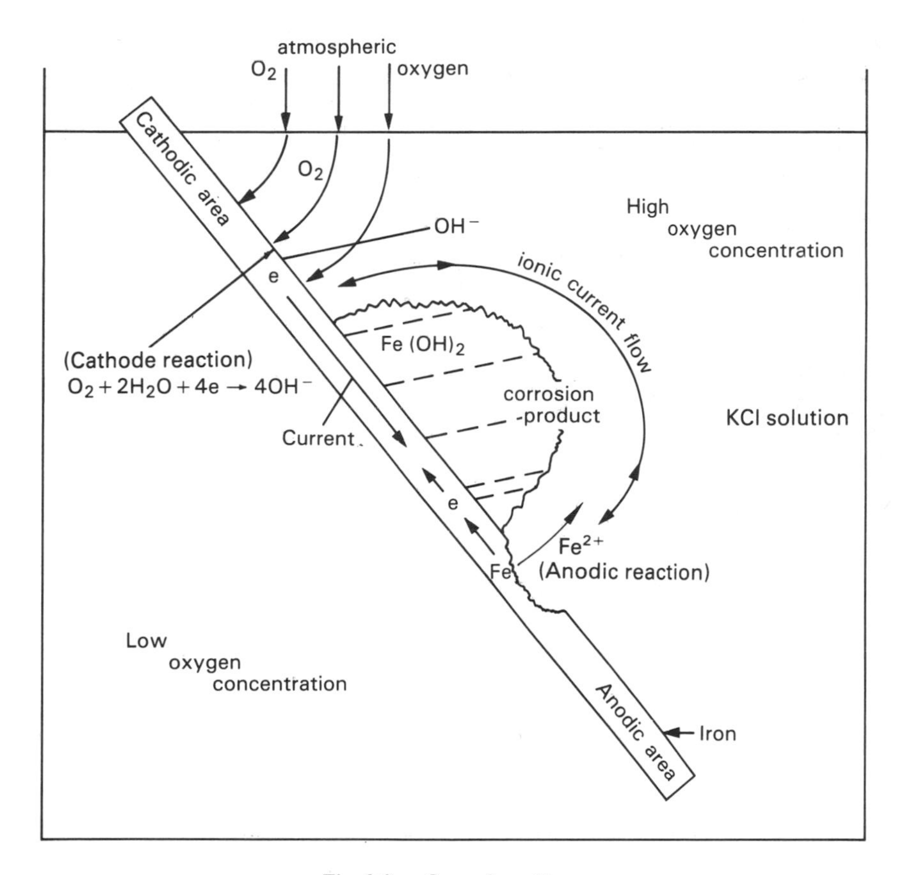

Fig. 3.3 — Corrosion of iron.

are apparent. A current flows within the metal from the oxygen-rich area near the surface of the solution to the oxygen-deficient areas lower down (the conventional current flow is in the opposite direction to the flow of electrons). The role of the chloride ion is important. Although it is not involved in the overall stochiometry of the reaction it raises the ionic conductivity of the solution and so assists in carrying the external current. The chloride ion is also particularly efficient in destabilising the oxide coating and stimulating dissolution of the metal. Other anions such as sulphate, nitrate, bromide and fluoride have a similar action but none of them is as effective as the chloride ion. This reactivity, together with the ubiquitous occurrence of chlorides, ensures that the chloride ion is high on the list of suspects when untoward corrosion is encountered — but this theme will be taken up in later chapters.

MECHANISM OF ATMOSPHERIC CORROSION

The electrochemical mechanism which has been so successfully established for the corrosion of metals in solution is difficult to demonstrate and investigate in atmospheric corrosion since the film of electrolyte is so thin but the deductive evidence is overwhelming. The corrosion process in a drop of sea water on a metal surface may be considered as part way to atmospheric corrosion and this process has been investigated in detail and the mechanism shown in Fig. 3.4 established.

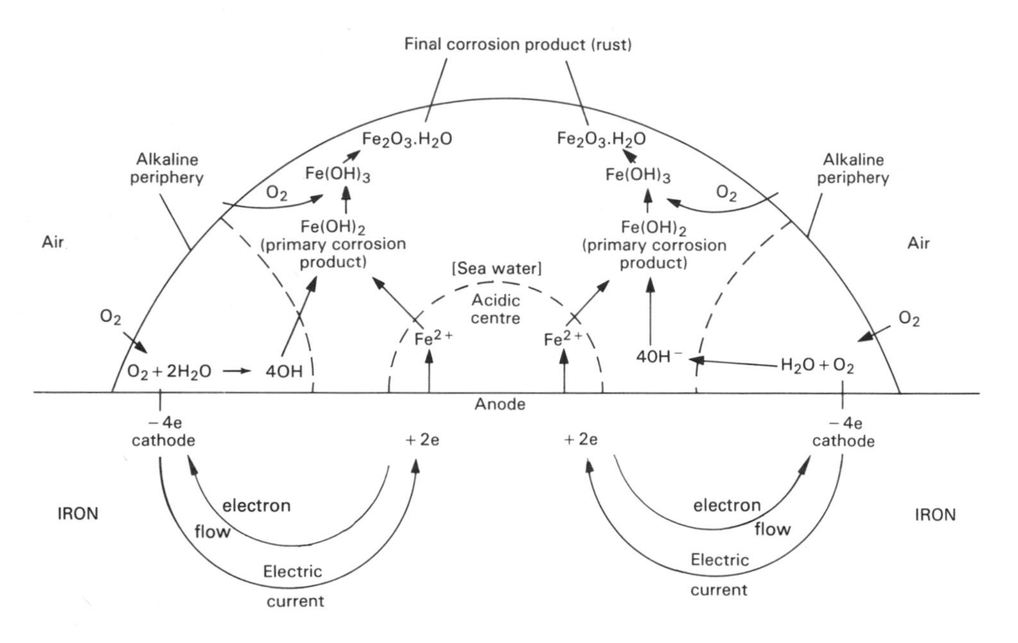

Fig. 3.4 — Corrosion reactions in a drop of sea water on iron.

The existence of the corrosion currents in the drop of salt water as shown in Fig. 3.4 was elegantly demonstrated by Blaha (1950) by applying a strong magnetic field vertically across the drop, when the solution rotated around the centre of the drop. The effect is that of an electric motor; the magnetic flux of the applied field reacts with the current in the solution caused by the flow of charged ions and so the conducting solution moves at right angles to the direction of the current flow and to the applied field. When the polarity of the applied field was reversed the direction of rotation reversed.

The electrochemical mechanism of atmospheric corrosion was studied by Rozenfel'd (1962) using multilayer sandwiches of metal films separated by thin layers of mica to provide electrical insulation. The surface at right angles to the layers of the sandwich was polished and thin films of electrolyte were spread over the surface and the potential differences and the current flows were examined and related to the corrosion effects observed. Other workers have further developed this approach and have produced senstitive electrochemical monitors for atmospheric corrosion. Gonzalez, Otero, Cabanas & Bastidas (1984), for instance, have produced a multilamellar cell, with an insulated central foil as a reference electrode. Under controlled conditions representative of atmospheric corrosion, the corrosion current determined from the electrochemical measurements was proportional to the observed corrosion.

The experimental evidence and the parallels between corrosion in air and in solutions, together with the success of deductions which assume an electrochemical mechanism and the practical observation that films of solution are present on corroding surfaces, leaves no doubt that atmospheric corrosion occurs by an electrochemical mechanism.

The electrochemical mechanism which gives such useful insights into the phenomenon observed in atmospheric corrosion has been largely established and studied through experiments on specimens immersed in solution. Although there is a very close similarity which allows mechanisms to be deduced and so gives insights into the phenomenon, the differences are also important and the difficulties of carrying out direct measurements on surfaces corroding in the atmosphere means that quantitative theoretical interpretations are seldom possible. Care and discrimination are needed in applying the basic electrochemical principles to the interpretation of corrosion in the atmosphere; to a large extent quantitative results for atmospheric corrosion rates can only be arrived at empirically.

The conditions of atmospheric corrosion give rise to variations in the way in which corrosion progresses when compared to immersed conditions: on a metal surface exposed to the atmosphere there are only limited quantities of water and dissolved ions but an unlimited access to oxygen from the air and the corrosion product must remain on the surface close to the point at which it is formed. Chemically stable corrosion products may form and if they are insoluble they provide a physical barrier, as for instance does the coating on copper when the metal reacts with chloride and sulphate ions to give a protective patina, as also does the insoluble lead sulphate which develops on lead exposed to industrial atmospheres. The corrosion of copper and lead are soon stifled in this way so that they are little attacked in urban atmospheres once these insoluble coatings have formed.

In contrast the ferrous salts which form on iron surfaces are soluble but the ferrous ions in solution are soon further oxidised by atmospheric oxygen to ferric ions and the insoluble hydrated ferric oxides are formed as the familiar rust. The rust is therefore formed away from the corroding surface and the soluble sulphate ion remains in solution so that it can continue to cause further corrosion. The sulphate ions remain close to the corroding surface which becomes uneven and pitted so that it is difficult to halt attack by physical cleaning. Even the ferric ions in the rust assist corrosion since they provide an adjacent and convenient oxidising agent for cathodic reaction by being reduced to the intermediate ferroso-ferric oxide. This reduction of the ferric oxide occurs in cycles which in turn are prompted by changes in temperature and humidity. The overall effect is to produce layers of porous rust which provide little or no protection.

REFERENCES

Appleby, A. J. & Mayne, J. E. O. (1967) *J. Oil Col. Chem. Ass.* **50,** 897.
Blaha, F. (1950) *Nature (Lond.)* **166,** 607.
Donovan, P. D. & Heron, J. T. (1971) Unpublished report.
Gonzalez, J. A., Otero, E., Cabanas, C. & Bastidas, J. M. (1984) *Br. Corros. J.* **19,** 2, 89.
Mayne, J. E. O. (1970) *Br. Corros. J.* **5,** 106.
Pourbaix, M. (1966) *Atlas of Electrochemical Equilibria in Aqueous Solutions.* Pergamon Press. Cebelcor.
Rosenfel'd, I. L. (1962) *Proc. 1st Int. Corros. Cong. (London),* Butterworths.

4

Atmospheric Corrosion

In considering atmospheric corrosion, steel must take pride of place on almost every count: certainly in terms of value, tonnage of metal exposed, tonnage returned to its native condition by corrosion, industrial importance and in the depth and extent of the studies carried out. Much of the work on the effects of corrosion on zinc, cadmium, nickel, chromium, and tin has been undertaken on thin coatings on steel since they are frequently used for the protection of steel. Aluminium is second to steel in intensity of study, with lesser study of atmospheric effects on copper, lead, magnesium and titanium.

CAUSES OF ATMOSPHERIC CORROSION

The classical work of Vernon (1935) established the major contributory effects of relative humidity, sulphur dioxide (the chief gaseous pollutant in industrial atmospheres), ammonium salts and particles of charcoal (representing the soot in industrial environments) in the atmospheric rusting of steel. Fig. 4.1 from Vernon's original paper clearly shows the role of these pollutants in conjunction with high humidities in causing corrosion.

Fig. 4.1 shows the weight increment with time of iron panels exposed in an environment of increasing humidity. The effects of a range of solid contaminants on the metal surface with 0.01% sulphur dioxide in the atmosphere, and of charcoal particles and ammonium sulphate without sulphur dioxide, were studied; the weight increment on a panel exposed continuously in an atmosphere at 99% RH containing 0.01% sulphur dioxide is also shown. Corrosion, where it occurred, began when the relative humidity reached a critical level which in this instance was about 80%, but the precise level of this critical humidity varies with both the type of contaminant present and with the composition of the metal. Vernon in fact distinguished two stages of critical humidity: the primary, where corrosion

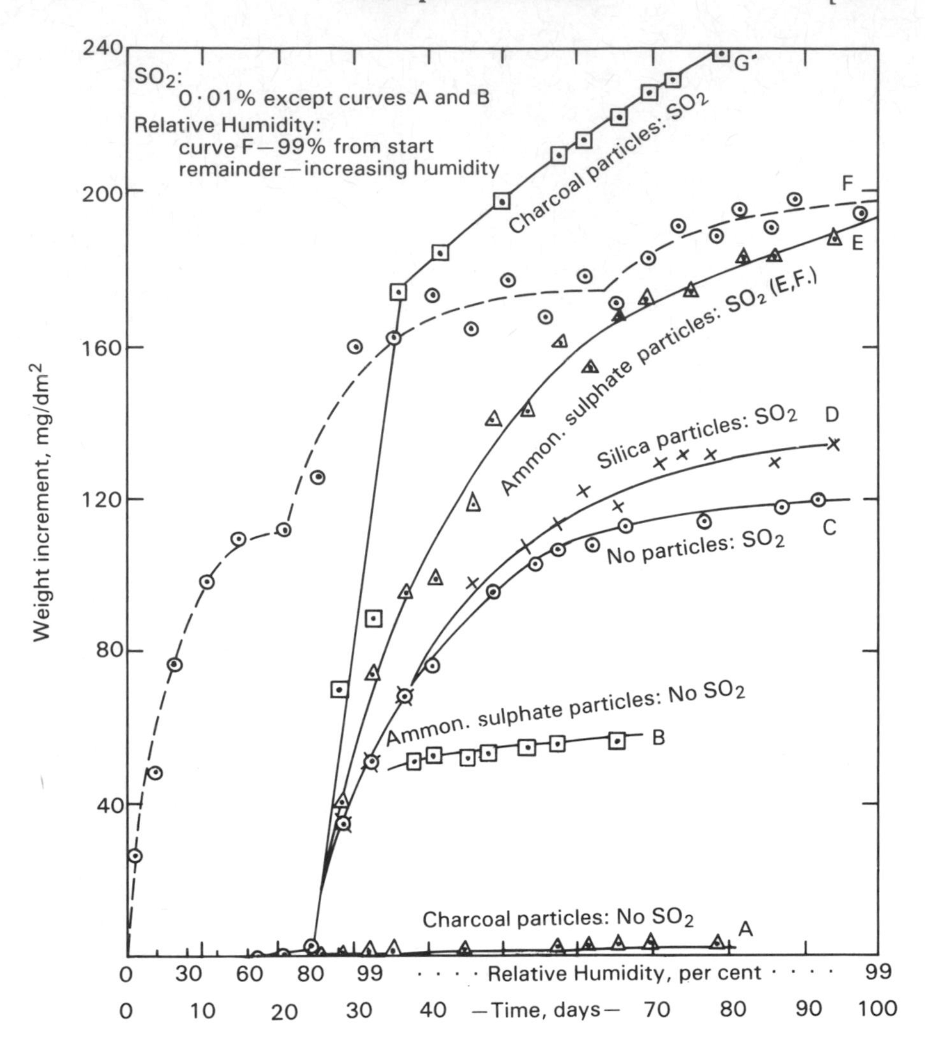

Fig. 4.1 — Effects of relative humidity and pollutants on corrosion of iron (from Vernon (1935); reproduced by permission of The Royal Society of Chemistry).

first begán, between 60 and70% RH, and the secondary, the onset of rapid corrosion at 80% RH. The critical humidity is considerably lower if marine salts are present. Chandler (1966) showed that in marine conditions measurable corrosion occurred at 40% RH, and Evans & Taylor (1974) showed that corrosion occurred on steel surfaces contaminated with sea salt down to, and even below, 35% RH.

At humidities above the critical level the highest rates of corrosion in Vernon's experiment were with sulphur dioxide in conjunction with charcoal. Ammonium salts gave lower rates of rusting than did sulphur dioxide alone. Inert silica particles had little effect on corrosion induced by sulphur dioxide and, although charcoal speeded corrosion when sulphur dioxide was present, on its own it had no effect.

Vernon confirmed the important role of particles on the atmospheric rusting of iron by exposing two samples of steel in an industrial atmosphere; one of these he protected with a muslin screen, sufficiently fine to keep out most particles. The sample behind the screen remained practically unattacked while the fully exposed specimen corroded.

The contaminants chosen by Vernon for his experiments corresponded to the impurities found in industrial atmospheres at that time. In the sixty intervening years the changes from extensive coal burning in domestic dwellings, small factories and railway engines, to the large-scale production of electricity for heating and power and the changes in scale of industrial processing have contributed to a dramatic reduction in the fall-out of solids in industrial areas (Royal Commission, 1971). The total emission of sulphur dioxide however has remained nearly constant at about 6 million tons over the United Kingdom but it is now more widely distributed. Over the same period the number of cars and lorries has increased dramatically and they distribute increasing amounts of oxides of nitrogen and combustion residues over wide areas (McColl, 1982). The role of these exhaust gases in causing corrosion has not been examined in any detail but they increase the level of ammonium salts, soot, organic acids and nitrates, all of which stimulate corrosion reactions.

Walton, Johnston & Wood (1982a) have simulated atmospheric corrosion of zinc, aluminium and iron, and studied rusts from a range of rural, industrial and marine sites. The most aggressive contaminants were salt, salt/sand, ash from iron smelters, municipal incinerator plume ash smuts and coal mine dusts. They concluded that the role of particulate matter is important and needs greater attention. The same workers (1982b) have constructed a test cell equipment and have described direct visual and photomicroscopical observation of atmospheric corrosion *in situ*. They studied the role of temperature, relative humidity, sulphur dioxide and particulate matter. Their work deserves study in the original papers; one major observation was the importance of flow rate and air flow pattern over the metal surface, which govern the rate of transfer of contaminants to the surface, and are at least as critical as the concentration of contaminants in the air in determining the rate of corrosion.

Studies on the chemistry of the sulphur dioxide released into the atmosphere reported by McColl (1982) suggest that much of it is converted to sulphur trioxide in the upper atmosphere. The sulphur dioxide and sulphur trioxide dissolve to give sulphurous and sulphuric acids. Any remaining sulphurous acid oxidises to sulphuric acid on the metal surface, possibly by reacting with the rust (see for instance: Kaneko & Inouye (1981)). In summary the reactions are:

$$SO_3 + H_2O \rightarrow H_2SO_4 \text{ (sulphuric acid)}$$
$$SO_2 + H_2O \rightarrow H_2SO_3 \text{ (sulphurous acid)}$$

$$SO_2+2Fe_2O_3 \rightarrow FeSO_4+Fe_3O_4$$

The ferrous sulphate formed through reaction between the metal surface and the sulphuric acid solution accelerates corrosion through several mechanisms. It increases the amount of water absorbed onto the surface when the relative humidity exceeds that at which hydrates of ferrous sulphate form; this humidity corresponds to the critical humidity observed in Vernon's experiments. The ferrous sulphate solution provides the electrolyte to carry the external corrosion current and the sulphate has a specific role in assisting anodic dissolution by becoming absorbed within the oxide film to give weakened, porous or conducting areas which allow the ferrous ions to pass more readily into solution. Analysis of the rust on corroding surfaces has shown that the sulphate tends to concentrate near the metal surface as sulphate 'nests' (see for instance Barton, Kuchynka, Bartonova & Beranek (1971)). Once rusting has started, removal of sulphur dioxide from the surrounding air does not stop corrosion, since the sulphate present close to the metal surface is involved in a cycle of reactions in which the soluble sulphate ions are regenerated. Cleaning steel by removal of the loose outer layers of rust leaves much of the active contaminant adhering to the metal so that rusting continues.

Active particles provide a catalytic surface for the cathodic reaction (the reduction of oxygen) and are instrumental in absorbing and concentrating contaminants such as sulphur dioxide and in retaining the film of water on the metal surface. They may also hinder repair of the oxide film so that anodic dissolution proceeds more readily.

Iron corrodes faster than any of the other engineering metals when fully exposed in industrial or marine atmospheres. Its comparatively lowly position in the electrochemical series (Table 3.1) below zinc and just above cadmium, and the passive range of stability of its oxide, which although restricted is similar to those of zinc and cadmium, give no indication as to why its susceptibility to corrosion should be so much greater then these two metals. Zinc and cadmium are in comparison so resistant to corrosion that they are frequently used as protective coatings on steel. The reason lies, at least in large measure, in the existence of the two valency states, the ferrous (Fe^{2+}) and ferric (Fe^{3+}), and the insolubility of the ferric hydroxide in near neutral solutions. This gives a ready mechanism for soluble ferrous salts to be converted to an insoluble and non-protective layer and at the same time to regenerate active sulphate ions which continue to play their essential role in further corrosion.

The overall picture of rusting as built up from Vernon's experiments and later work by Evans (1972), and Evans & Taylor (1974), is illustrated in Fig. 4.2.

The build-up of rust on mild steel gives a slight degree of protection, reducing the rate of corrosion by about 50%. In heavily polluted industrial atmospheres the initial rate of attack of about 0.15–0.20 mm a year over the first one to two years falls to a relatively steady rate of about 0.10 mm a year on fully exposed surfaces but attack is often uneven and serious pitting may occur.

ALLOY STEELS

The rate of rusting of steel is decreased by additions of up to 0.15% of copper. Small additions of chromium, nickel, manganese and aluminium also decrease the rate of rusting, but their effects are less predictable and less marked. The corrosion rate of copper-rich steels is similar to mild steel for the first year but after approximately two years falls to about 0.03 mm a year in industrial atmospheres in the United Kingdom. Low-alloy slow-rusting steels have been developed by optimising additions of these elements for use in industrial atmospheres. These steels are intended to be used without painting or other protectives (see Schmitt & Gallagher, 1969) but there remain some reservations on their use in marine environments, in heavily contaminated industrial areas, and where contamination from de-icing salts used on roadways is likely. The corrosive effects of exposure on copper steel and on Cor-Ten slow-rusting steel in an urban atmosphere, taken from the results of Larrabee and Coburn (1962) are shown in Table 4.1.

Additions of over 12% of chromium to steel produce the range of corrosion-resisting 'stainless steels' in which the passivating effect of chromium confers protection under many but not all conditions. Crevice corrosion, intergranular corrosion and stress corrosion are particular exceptions referred to later.

NON-FERROUS METALS

The atmospheric corrosion rates of non-ferrous metals are generally lower than those of steel. Table 4.2 contains results from a series of exposures of pure metals and some alloys, carried out by Clarke & Longhurst in Nigeria (1962), in three representative environments: jungle, town and surf beach.

Although the results in Table 4.2 are from exposures in the tropics they are similar to those recorded in temperate areas. The marine site was near a surf beach but 200 yards above the high water mark so that the level of contamination by marine salt was limited; the urban site was 25 miles from the sea with moderate levels of sulphurous pollution from a rail depot, power station and dwellings in the vicinity. The jungle site was 35 miles from the open sea, and 12 miles from the nearest town although it was down-wind of it so that pollution would be expected to be above that in a remote jungle but nevertheless the corrosion rates are among the lowest recorded for outdoor exposure in other than arid areas.

The highest corrosion rates in Table 4.2 are of steel in polluted environments, with the corrosion rate at the coast being about three times as great as that in the town. By comparison the attack on all the non-ferrous metals was low. Magnesium alloy, brass, copper, zinc, tin, tin–zinc alloy, and LM-4 aluminium alloy showed a measurable level of corrosion in the marine atmosphere but the most rapidly corroding metal of these, magnesium,

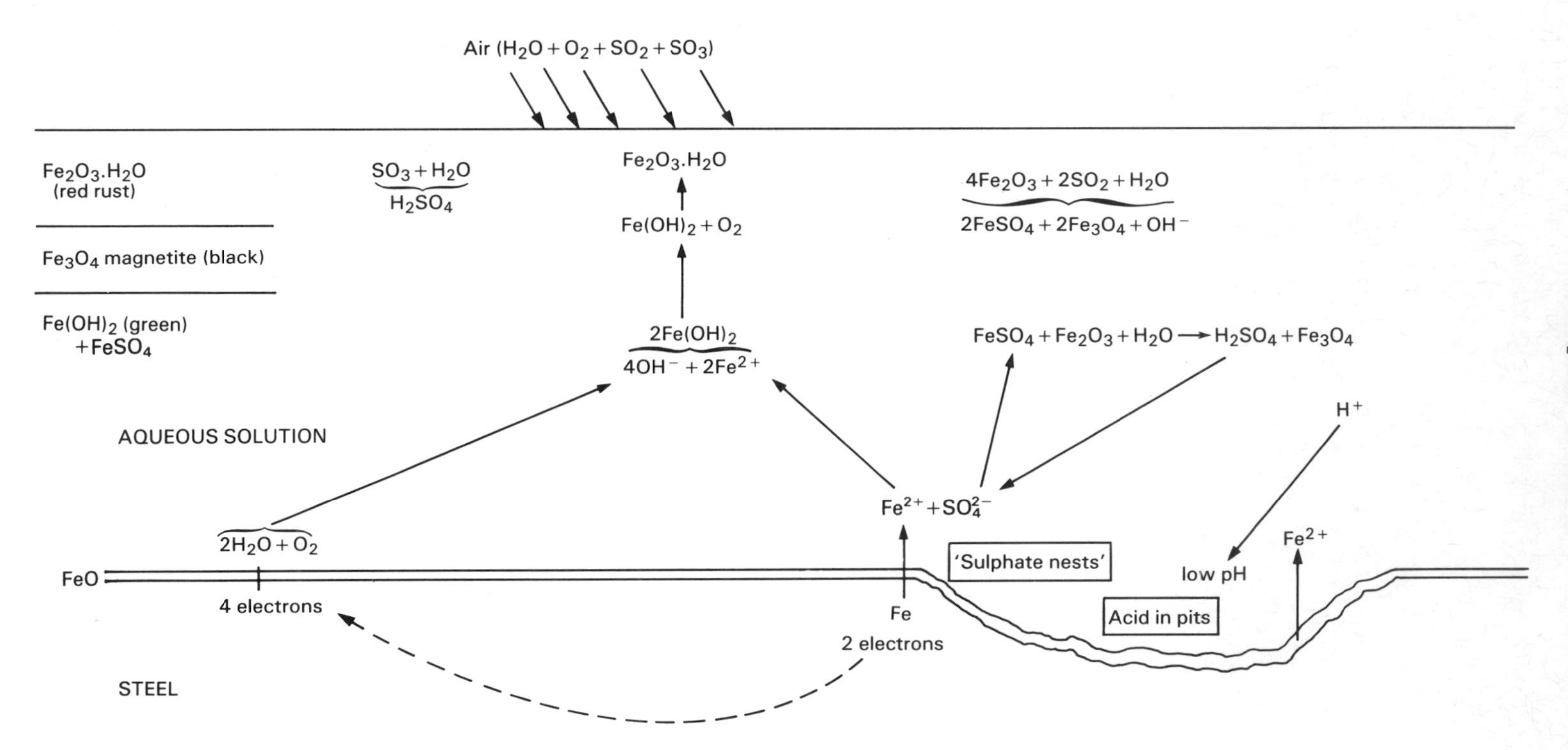

Fig. 4.2 — Reactions in the atmospheric rusting of steel.

Table 4.1 — Corrosion of low-alloy steels in an industrial atmosphere (Kearney, New Jersey, USA) (reproduced with permission from Larrabee & Coburn (1962)).

	% Alloying elements								Depth of attack[a] (μm)						
									years						
	C	Mn	P	S	Si	Cu	Ni	Cr	1.0	1.5	2.5	3.5	10	15.5	20
Structural carbon steel	0.17	0.57	0.019	0.05	0.043	0.05	0.02	0.02	59		79		123		163
Structural copper steel	0.18	0.49	0.024	0.034	0.025	0.32	0.02	0.02	30		39		63		88
Cor-Ten steel	0.06	0.48	0.11	0.030	0.54	0.41	0.51	1.0		15		18		15	

a Mean of attack on upper and lower surfaces.

Table 4.2 — Corrosion rates of metals in tropical atmospheres (reproduced with permission from Clarke & Longhurst, 1962).

	Corrosion (μm/year)[a]		
Metal	Jungle	Town	Coast
Mild steel	4	42	120
Magnesium alloy	17		13
Copper	0.5	2	2
Brass	0.7		4
Zinc	0.5	1	3
Cadmium	0.4	1	0.3
Tin–Zinc (80:20)	0.5	1	3
Tin	0.2	1	3
Lead	0.4	1	0.3
Nickel	0.03	0.1	0.2
Aluminium	<0.03		<0.2
Aluminium alloys			
BS 1470 NS5	<0.03		<0.08
BS 1470 HS15	0.04		0.1
BS 1490 LM4-M	<0.03		3

a Averaged over front and rear surfaces of 150 mm×100 mm panels exposed for two years at 45°.

corroded at only one-tenth the rate of steel while the other metals corroded at only 3–4% of the rate recorded for steel. Cadmium, lead, nickel and most of the aluminium alloys tested corroded very slowly, all giving rates of corrosion below 0.3% of that of steel.

The inference from the outdoor exposures in Table 4.2 might be that steel corrodes rapidly outdoors whereas non-ferrous metals suffer little corrosion and do not need protection. There is a measure of validity in this inference; steel usually needs protection, and aluminium, lead and brass for instance are often used without protection while zinc and cadmium coatings are used to protect steel outdoors.

INTENSIFIED CORROSION

There are however conditions under which non-ferrous metals fail rapidly and sometimes catastrophically through corrosion, and steel itself can fail through corrosion much more rapidly than the rates of corrosion in Tables 4.1 and 4.2 suggest. Rapid corrosion may result when high levels of contaminants accumulate, as for instance happens on the undersides of motor cars in damp climates when de-icing salt is used; the dirt also present retains water and accelerates attack. Sometimes rapid failures occur through selective attack at grain boundaries, crevices or along axes of high deformation. In some enclosed atmospheres particular contaminants are encountered which and to very rapid attack of some metals. In an investigation carried out by Donovan & Moynehan (1965) for instance, enclosure of zinc at high humidity with an acrylic paint for three weeks resulted in an attack of 0.05 mm, equivalent to the total attack expected in 10 years in a marine environment.

Any assessment of corrosion hazards must take account of the particular conditions of metallurgy, design, stress and surface condition which so often accelerate or localise corrosion. Many catastrophic failures of metallic structures, on which little or no corrosion products are visible, are nevertheless caused by corrosion, and in many others corrosion is a contributory factor. The importance of corrosion as a contributory factor in many so-called mechanical failures is difficult to exaggerate — in the writer's experience corrosion makes a significant contribution in at least 70% of unexpected 'mechanical' failures of metals (it is worth noting, although beyond the scope of the present book, that many mechanical failures of plastics are similarly caused, or hastened by reactions with the environment).

The major factors which modify the corrosion process in atmospheric conditions are:

— aspects of design, assembly and use. The effects occur especially at joints, contacts and from drainage waters, giving crevice corrosion, attack through oxygen shielding, bimetallic corrosion and weld decay;
— alloying and metallurgical condition;
— the presence of strain, which is associated with stress corrosion, corrosion fatigue, hydrogen embrittlement and whisker growth;
— vibration (or fretting between adjacent surfaces) to give fretting corrosion and fretting fatigue;
— sources of organic acid vapours within enclosed environments;
— contact with contaminated materials, especially those containing chloride residues;
— the effects of contaminants already on the metal surface;
— rodents, insects, termites and micro-organisms.

The next chapter (Chapter 5) descibes and discusses the first four of these. The next three are particularly important in transport and storage and each is discussed in separate chapters, respectively Chapters 6, 7 and 8. Biological processes of deterioration extend beyond the scope of this book but the major effects are summarised in Chapter 9.

REFERENCES

Barton, K., Kuchynka, D., Bartonova, Z. & Beranek, E. (1971) *Corros. Sci.* **11,** 937.
Chandler, K. A. (1966) *Br. Corros. J.* **1,** 264.

Clarke, S. G. & Longhurst, E. E. (1962) *Proc. 1st Int. Corros. Cong. (London)*. Butterworths.
Donovan, P. D. & Moynehan, T. M. (1965) *Corros. Sci.* **5,** 803.
Evans, U. R. (1972) *Br. Corros. J.* **7,** 10.
Evans, U. R. & Taylor, C. A. J. (1974) *Br. Corros. J.* **9,** 26.
Kaneko, K. & Inouye, K. (1981) *Corros. Sci.* **21,** 9/10, 639.
Larrabee, C. P. & Coburn, S. K. (1962) *Proc. 1st Int. Corros. Cong. (London)*. Butterworths.
McColl, J. G. (1982) Keith, L. H. (Ed.) *Energy & Environmental Chemistry*, Vol. 2. Butterworths.
Royal Commission on Environmental Pollution. First Report (1971).
Schmitt, R. J. & Gallagher, W. P. (1969) *Materials Protection* Dec. 70.
Walton, J. R., Johnston, J. B. & Wood, G. C. (1982a) *Br. Corros. J.* **17,** 2, 65.
Walton, J. R., Johnston, J. B. & Wood, G. C. (1982b) *Br. Corros. J.* **17,** 2, 59.
Vernon, W. H. J. (1935) *Trans. Faraday Soc.* **31,** 1, 668.

FURTHER READING

Ailor, W. H. (Ed.) (1982) *Atmospheric Corrosion.* John Wiley.

5

Non-uniform corrosion and conjoint action

This chapter contains a summary account of a range of corrosion processes which are phenomenologically distinct and are therefore readily considered under individual headings, and yet many of them share characteristics and mechanisms. The processes are described separately, as this eases both description and understanding, but in an order which has been selected to allow the features which many of these processes share to be more readily appreciated and to avoid needless repetition.

Conjoint action refers to the failure mechanism which results when chemical and physical stresses combine to give more rapid failure than would be expected by their simple additive effects. The types of corrosion described mostly fit within the scope indicated by the title of this chapter but for completeness some physical processes which can only loosely be described as corrosion are included.

BIMETALLIC CORROSION

Bimetallic corrosion can occur when two metals are in direct electrical contact and are also connected by an external path through an electrolyte, so that the anodic and cathodic reactions can be partitioned between the two metals in a way that allows the anodic reaction to be concentrated on one of the metals. The net effect is to increase the rate of dissolution of the anodic member of the couple while the cathodic member is protected. The mechanism is essentially that illustrated in Fig. 3.1.

Bimetallic corrosion is often encountered when metals are immersed in electrolytes, typically on the hulls of ships, inside boilers and cooling water systems and in chemical plant. Under immersed conditions accelerated attack may occur on the anodic member of the couple at up to considerable distances from the contact, while the cathodic member is correspondingly protected away from the area of contact. The effect is often used to control corrosion by providing consumable anodic attachments, known as sacrificial anodes, so that the critical part of the structure is cathodic and does not suffer attack.

The film of electrolyte which is present on a metal surface exposed to an industrial atmosphere is thin and offers a high resistance to ionic currents so that the bimetallic effect is restricted to an area close to the contact; in practice this is seldom more than 2–3 mm.

A typical bimetallic couple is illustrated in Fig. 5.1. In this couple of zinc and copper, zinc is more electropositive and passes into solution relatively easily. Copper is less electropositive than zinc, and passivates more readily because of the wider range of stability of its oxide (see Table 3.2). It also has a lower overpotential and is a much more efficient catalyst for the cathodic reaction, the reduction of oxygen. The net result shown in Fig. 5.1 is concentration of the attack on the zinc close to the area of contact and some protection of the copper near the contact.

Copper, silver, carbon and the noble metals have low overpotentials while zinc, iron, aluminium and many of the more active metals have high overpotentials. Contact of these reactive metals with the former group gives especially intense bimetallic effects since the cathodic member provides extra area for the cathodic reaction and lower polarisation over this extra area.

The high resistance of the external path through the electrolyte restricts attack to the area very close to the contact. Because of this limitation bimetallic corrosion only gives serious effects on atmospheric exposure in three conditions:

— where the contact is frequently wetted with a strong electrolyte, typically on board ship, near the sea, or on the underside of cars or other items exposed to splash from roads where de-icing salt is used;
— where a more noble metal is deposited on a reactive metal by displacement:

$$\text{e.g } Zn + Cu^{2+} \rightarrow Zn^{2+} + Cu$$

which occurs where water which contains copper ions, after draining over a copper roof or a copper lightning conductor for instance, flows over a galvanized roof, in iron or aluminium guttering, or over aluminium panels or frames. Attack may be especially damaging as the copper is widely distributed over the other metal, and in intimate contact with it, so that the anodic and cathodic areas are very close to one another and the resistance to the external ionic current is low. Under these conditions copper again forms a very efficient cathode;
— where the couple is exposed to a contaminated environment and the geometry of the assembly gives small areas of an anodic metal in contact with large areas of a surrounding cathodic metal. Thus stainless steel or copper panels fastened with aluminium, or zinc-plated steel, bolts or rivets, are liable to rapid attack on the small fasteners.

An important example of this last effect is encountered with metal

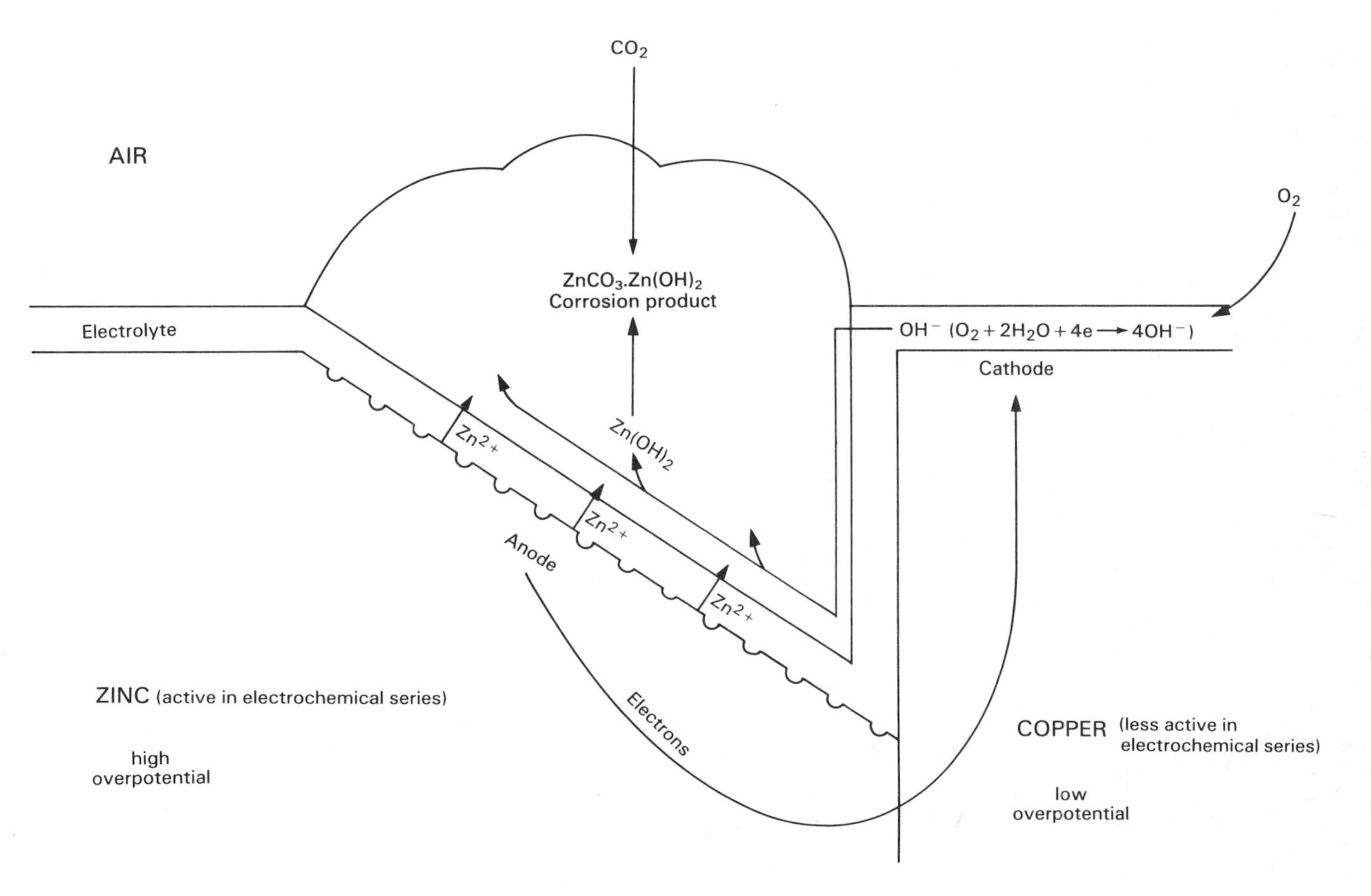

Fig. 5.1 — Bimetallic corrosion: accelerated attack on zinc.

coatings on metal substrates. When nickel and chromium coatings are used for instance to protect steel, rusting occurs at any pores or gaps in the coating since the steel base and the metal coating form a bimetallic couple, with the steel anodic to the surrounding cathodic coating so that a small steel anode is surrounded by a large nickel/chromium cathode. Conversely, although zinc, cadmium and aluminium protect steel initially by exclusion, they also provide protection at pores or other discontinuities since they are anodic to the steel and give sacrificial protection. These anodic metals are therefore preferred for corrosion protection of steel to coatings such as gold, copper, silver and nickel/chromium which are more resistant to corrosion than the anodic metals but give a lower standard of protection because the steel rusts rapidly at pores or sites of damage. Nickel is anodic to copper and brass and therefore nickel/chromium coatings give good protection on these metals.

The British Standards Institution *Commentary on Corrosion at Bimetallic Contacts and its Alleviation* (BS PD 6484: 1979) gives detailed and useful advice on all aspects of the subject. The Department of Industry's Guide No. 14 (1982) gives summary advice. Evans (1963) quotes from an advisory publication used in the design of US Naval aircraft. This divides metals into four groups:

— magnesium and its alloys;
— cadmium, zinc, aluminium and their alloys;
— iron, lead, tin and their alloys (other than stainless steel);
— copper, chromium, nickel, silver, gold, platinum, titanium cobalt, rhodium, and their alloys, and stainless steels and graphite.

No bimetallic effects are likely between metals in the same group; some effects are to be expected between metals in adjacent groups and serious effects are probable between metals from non-adjacent groups. This is an oversimplification since the tin-steel couple may accelerate corrosion of steel and some combinations from the final group may give bimetallic effects. The groups do give a useful start in considering possible contacts for atmospheric exposure, and would be unlikely to give problems in any but the most corrosive environments.

Protection from bimetallic corrosion may sometimes be obtained by metal coating one of the metals of the couple. Titanium for instance has often been plated with cadmium when in contact with aluminium (on high speed aircraft, where the outer skin may become hot, problems were encountered through liquid metal attack by the cadmium on the stressed titanium, but that is another problem). An alternative method of protection is to separate the metals by insulating materials. This technique is often applied to contacts of fasteners by using insulated washers and sleeving.

Many specifications and guides have used the electrochemical series of potentials as a basis for recommendations on the corrosion hazards from contact between different metals. When the fallacy of this approach was realised an alternative series was produced based on potentials of metal electrodes in aerated sea water and the series then did have some relevance to marine environments although even here there were anomalies, since although the potential may indicate which metal is initially anodic, once current flows, polarisation may cause a reversal of polarity. With other environments the use of such potentials has no valid basis, since it takes no account of polarisation and the reacting species from the environment.

However attractive the concept of a ranking series, reality demands a more informed approach since even to a first approximation several competing factors are involved and details of design, environment and protective treatments need to be taken into account. The British Standards Commentary (1979) should be consulted in any cases of doubt since it is a consensus of experts on what is still an area full of uncertainties and pitfalls. Johnson & Abbott (1974) have published results of bimetallic couples between mild steel and a wide range of aluminium, copper, nickel, iron, lead, and magnesium alloys; carbon, and cadmium, zinc, tin and chromium metals. These were exposed in immersed conditions, and in industrial, urban/rural and marine conditions. Their paper is a good source of information on bimetallic contacts with steel.

INTERGRANULAR OR INTERCRYSTALLINE CORROSION

Corrosion attack sometimes concentrates at boundaries between the grains of alloys, penetrating to considerable depths with little general attack on the surface so that the metal loses its strength and cohesion with little external sign of corrosion. The phenomenon is usually associated with alloys which have a minor phase precipitated at the grain boundaries. An active anodic path results either through this phase, or through the area of the grain immediately adjacent to it which has become depleted in the elements which have concentrated in the minor phase in its precipitation. Intergranular corrosion has been most frequently reported in copper and aluminium alloys, and in stainless steels, but is also encountered in iron and nickel alloys.

Resistant alloys have been developed both through modifications of composition and heat treatment cycles. These improved alloys have practically banished the incidence of this type of corrosion and when it is encountered, at least under non-immersed conditions, it almost certainly indicates an unwise departure from the recommended heat treatment. Aluminium/magnesium and aluminium/copper alloys after some heat treatments are susceptible to intercrystalline corrosion. In the former alloys the Al_3Mg_2 precipitate which forms at grain boundaries is anodic to the bulk of the metal and dissolves preferentially (see for instance Mazurkiewicz (1983)). In the copper-containing alloys the $CuAl_2$ precipitate is relatively inert and it is the depleted zone, low in copper, adjacent to the precipitate which dissolves; the $CuAl_2$ precipitate also has a lower overpotential than the rest of the grain so that the cathodic reaction is assisted and results in increased attack (see Mazurkiewicz & Piotrowski (1983)).

Intergranular corrosion of austenitic stainless steels is associated with precipitation of chromium-rich carbides. The zone adjacent to the precipitate is depleted in chromium; if the level of chromium in this area falls below about 12% the passivity of the metal in the depleted area is lost and it is liable to corrode. The susceptibility of stainless steels to integranular corrosion arises from heat treatments within the temperature range 550–850°C; the effect is illustrated in Fig. 5.2.

Above 850°C the chromium-rich carbide dissolves and remains in solution but may precipitate on cooling through the critical range; it does not form on heat treating at temperatures below 550°C. Susceptibility can be avoided by keeping the carbon level below 0.03%, by quenching rapidly through the critical temperature range, or by adding about 1% of niobium or titanium since these metals have high affinities for carbon and so prevent the chromium-rich carbide forming.

EXFOLIATION OR LAYER CORROSION

Some rolled aluminium alloys suffer intergranular corrosion but as the grains are flattened the effect is to cause exfoliation sometimes likened to the layered structure of flaky pastry. Aluminium/magnesium and aluminium/zinc/magnesium alloys which have been overaged, are liable to suffer from exfoliation. Fig. 5.3 shows a typical example. As with other forms of intergranular corrosion, exfoliation is liable to give dramatic decreases in strength before corrosion is visible.

Robinson (1983), in an investigation of the wedging stresses produced by the corrosion product in exfoliation corrosion of an aluminium alloy (L95 plate in hardened heat-treated condition), showed that the maximum stress was 1.5% of the yield stress (4.8 MNm^{-2}), but calculated that the stress intensity was sufficient to propagate cracking by stress corrosion. He concluded that exfoliation corrosion propagates by a stress corrosion mechanism (note: stress corrosion is discussed later in this chapter). In support of this theory he showed that propagation could be stopped by countering the internal tensile stress with applied compression.

CREVICE CORROSION

Water, with any dissolved contaminants and solid debris collects in crevices at junctions between metals or between metals and non-metals. Metals which depend on passivation for their resistance to corrosion are laible to corrode within crevices where the metal remains wet for longer periods than on exposed surfaces, contaminants concentrate and access of oxygen is restricted, so that once the oxide film is damaged anodic dissolution can continue as oxygen is not available to repair the film. From Fig. 5.4 it can be seen that at the mouth of the crevice the cathodic reduction of oxygen to hydroxyl ions gives an increase of pH.

Deeper in the crevice where the concentration of oxygen is low anodic dissolution of the metal occurs. At some intermediate point ferrous ions from the anodic areas combine with hydroxyl ions from the cathodic reactions and a precipitate of ferrous hydroxide forms. In the solution adjacent to the anodic areas the pH decreases since the mobility of the

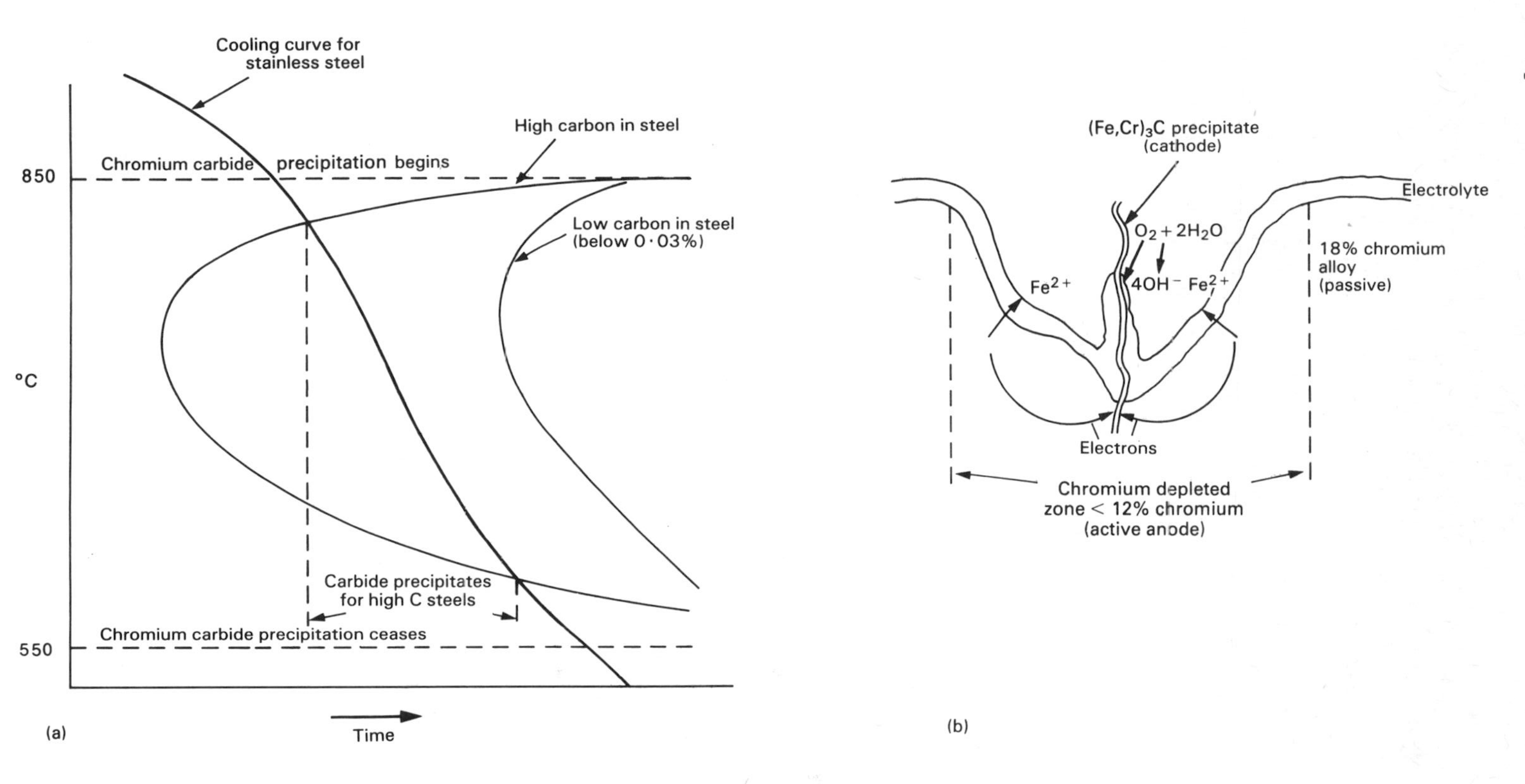

Fig. 5.2 — Intergranular corrosion of austenitic stainless steel. (a) Intergranular carbide precipitation. (b) Grain boundary attack of depleted zone.

Fig 5.3 — Layer corrosion of extruded aluminium alloy. (©Controller, Her Majesty's Stationery Office 1986.)

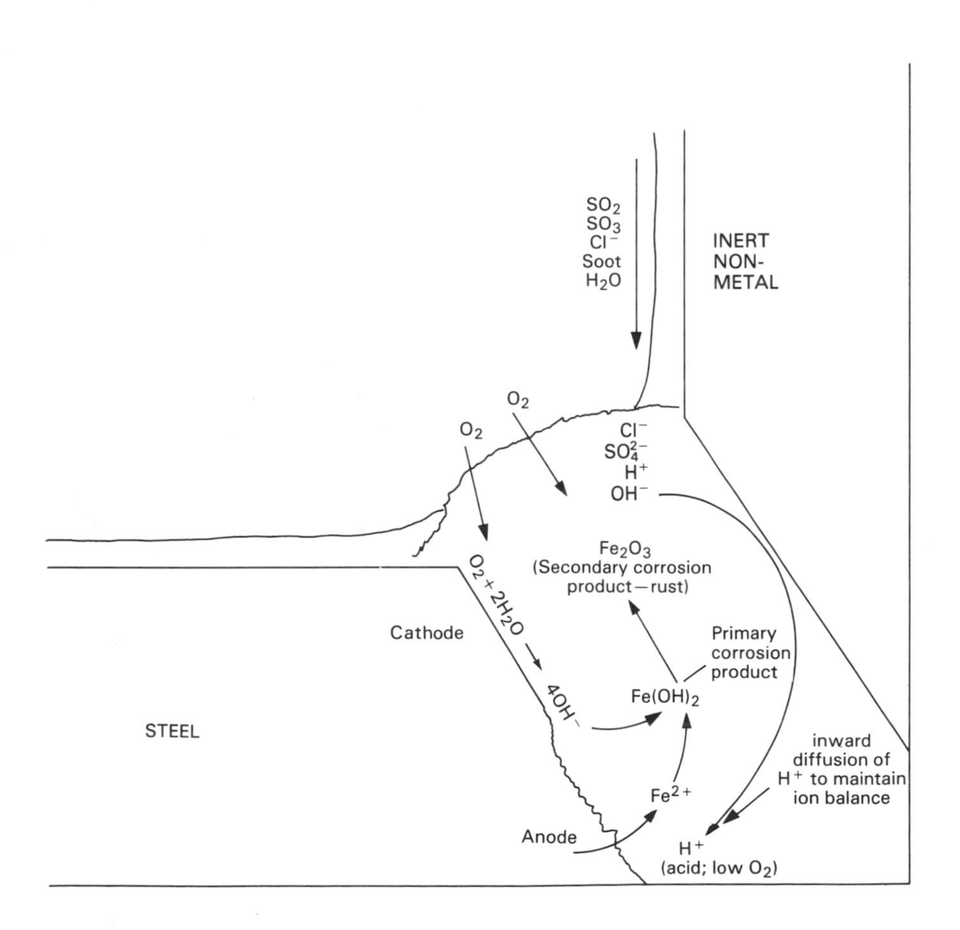

Fig. 5.4 — Crevice corrosion.

hydrogen ion is higher than that of any other ion and the hydrogen ion therefore diffuses preferentially towards the anodic areas to maintain the local balance of ions as positive ferrous ions are removed by precipitation. The acid condition which thus develops within the crevice stimulates further dissolution. The same effect can occur in bolted or welded structures.

WELD DECAY

Across brazed, welded or soldered joints a series of adjacent zones develop which differ in their chemical or metallurgical composition because of differences in temperature history or from mixing of the base metal with the brazing metal or solder. Sometimes a narrow band forms which is actively anodic to the areas on either side of it and in corrosive conditions preferential trenching occurs.

Welded stainless steels were at one time particularly susceptible to weld decay. Its occurrence is associated with precipitation of chromium carbide, and the level of carbon, chromium and the heat treatment regime are critical. Chromium carbide dissolves in steel above 850°C and remains in solution only if the steel is cooled rapidly, as shown in Fig. 5.2. In welding it is inevitable that part of the heat-affected zone (HAZ) alongside a weld bead will reach this critical condition and chromium carbide may be precipitated. The effect is illustrated in Fig. 5.5.

Weld decay of stainless steels can be avoided if the carbon content is kept below 0.03%, or if about 1% of niobium or titanium metal is added as a stabilizer. When stabilized austenitic stainless steels are welded, a zone immediately adjacent to the weld bead may become heated to a sufficiently high temperature for the titanium or niobium carbide to dissolve and the chromium-rich carbide may then precipitate as the alloy cools through the critical temperature range. Selective attack is then liable to occur along a narrow band adjacent to the weld bead, when it is known as knife line attack. In contrast weld decay occurs a few millimetres away from the weld bead and spreads over a wider band.

Under atmospheric conditions failures of welded, brazed and soldered joints are likely to be associated with corrosion by flux residues, or in crevices, or through physical failure at dry joints or at brittle phases which have formed, and these should not be confused with weld decay or other types of selective attack.

STRESS CORROSION

The combined effect of stress and corrosion sometimes produces cracking of metals at levels of stress well below the tensile strength of the metal. The cracking has the characteristics of brittle fracture and there is little or no general corrosion on the exposed surfaces. Stress corrosion is often associated with specific combinations of contaminant and alloy.

The effect was first recognised in the cracking of British army brass cartridge cases in India. As failures were particularly prevalent in the rainy

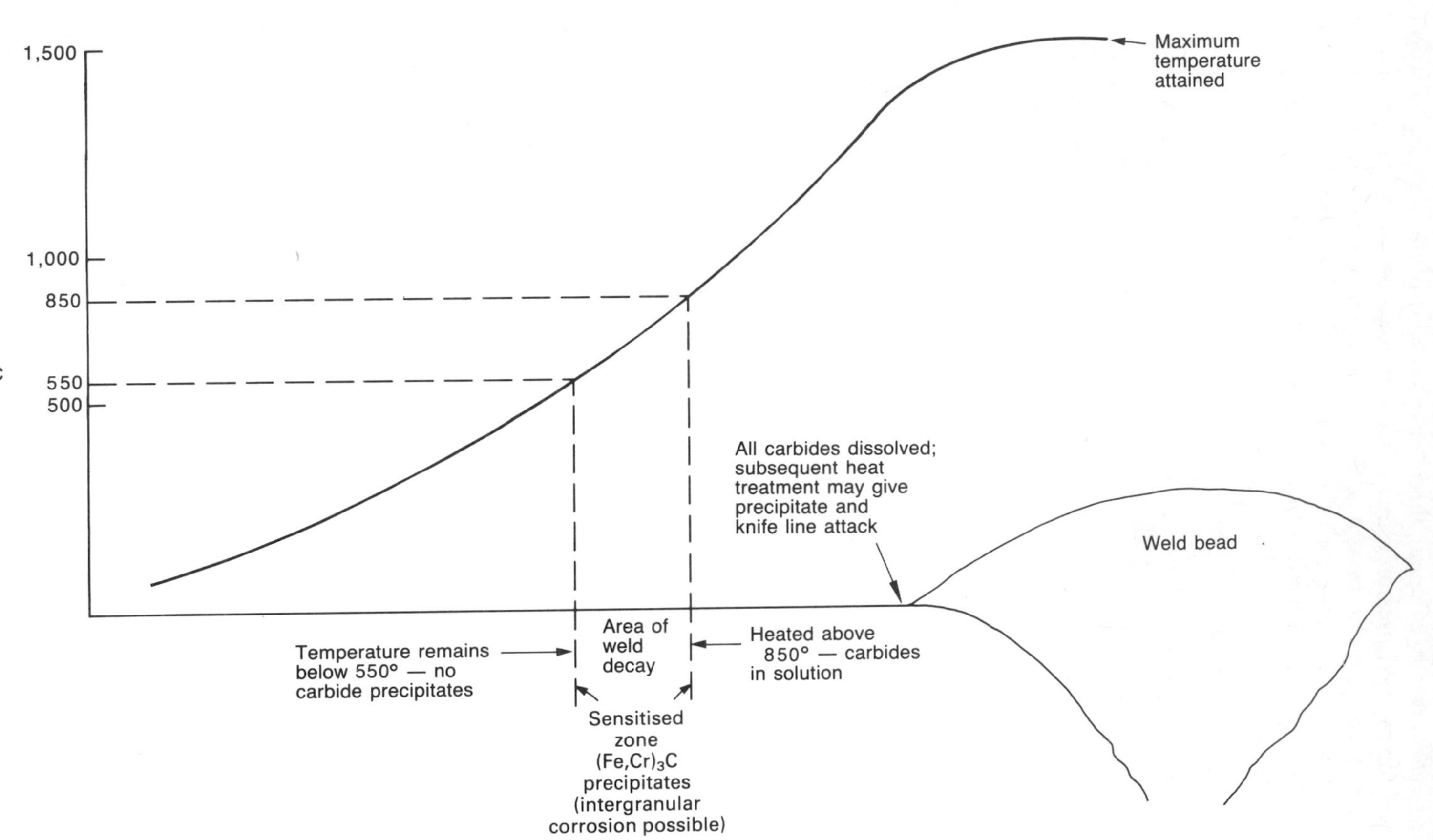

Fig. 5.5 — Welding-induced sensitivity to weld decay and knife line attack of stainless steels.

season the phenomenon became known as season cracking. Investigations showed that failure could be induced by exposure of assembled rounds over dilute solutions of ammonia. The stresses were derived from both the forced fit of the projectile into the case and from the hoop stresses introduced during the deep drawing process of manufacture. A typical failure of a deep drawn brass cup is shown in Fig. 5.6; in this instance the stress was the

Fig. 5.6 — Stress corrosion cracking of deep drawn brass (70:30) cup. (©Controller, Her Majesty's Stationery Office 1986.)

residual hoop stress remaining after manufacture.

The internal stresses in 70:30 brass may be relieved by anncaling for 1 hour at 275°C, and the failure shown in Fig. 5.6 occurred because the stress-relieving heat treatment had been omitted. Since the first failures with brass were encountered other examples of stress corrosion cracking have been identified. These include: mild steel boilers containing water and steam at high pressures; high strength aluminium alloys in aircraft components and high pressure gas cylinders; austenitic stainless steel in chemical plant; high strength precipitation hardened martensitic and semi-austenitic stainless-steel springs; high strength low allow steel pressurised gas pipelines; titanium alloy rocket motor cases in contact with dinitrogen tetroxide, and certain strong magnesium alloys. The list is far from a complete record of all failures and what was at one time thought to be an unusual phenomenon which required a very specific set of conditions closely defined by stress level, metallurgical conditions, composition of contaminant and temperature, has proved to be of more general occurrence.

It is unfortunate that the greatest incidence of stress corrosion cracking is among high strength alloys having intrinsic resistance to corrosion, and under heavy stress, just the conditions where high cost materials are likely to be fulfilling a difficult and vital role.

The damaging stress may be from external loading but may also arise from forced fit of components, damage or distortion, or be present as internal stress induced during manufacture and processing; hoop stresses in cylindrical structures are often particularly high. The hazard is increased with welded items since a wide range of metallurgical conditions extend across the weld and the neighbouring heat-affected zone. High internal stresses are often also associated with welds. Because of their importance as a method of fabrication, welding procedures have been developed for avoiding susceptibility to stress corrosion but care is needed to ensure that only proven procedures are followed.

Four stages may be distinguished in the process of stress corrosion: incubation, crack initiation, crack propagation, and final mechanical failure when the critical stress intensity is exceeded. The first two of these occupy the major proportion of time to failure. The time taken for cracks to propagate varies from minutes to months but it is only at this stage that there is sufficient evidence to detect that stress corrosion is occurring. Searching for such cracks is often a major task in the routine testing of airframes, and this precaution sometimes allows susceptible materials to be used in aircraft components but in few other industries is this level of inspection possible and usually the first sign of stress corrosion is a significant structural failure. Examination of the fracture surfaces after failure usually reveals the characteristics of brittle failure in the area over which the crack has progressed by stress corrosion. The cracking may follow grain boundaries (intergranular) or may propagate across grains (transgranular).

Incubation is an indeterminate but essential stage during which the conditions for crack initiation are established. It may involve thinning or thickening of the surface film, concentration of contaminants over part of the surface, exhaustion of oxygen in the solution adjacent to the metal surface and the formation of pits which may act as stress concentrators and may also give local changes in environmental conditions, and movements, or pinning, of dislocations, sometimes with consequent slip steps emerging on the surface. The interaction of these slip steps with the surface film to give rupture in lines of weakness where attack may begin, appears to be important in some instances such as the cracking of brasses. It is certainly significant that most susceptible materials form protective films, and that conditions near active/passive transitions are conducive to stress corrosion; incubation in most instances appears to involve subtle but critical changes in the surface film. With high strength steels, incubation may entail pitting, and then a local increase in hydrogen ion concentration, with consequent reactions releasing hydrogen on the metal surface which is absorbed into the steel to cause cracking through hydrogen embrittlement.

Once incubation has produced the appropriate conditions for selective dissolution, through localised chemical/mechanical interaction, then cracking commences and in susceptible metals continues, either as branched cracks or as single cracks, until mechanical failure occurs. In solution propagation can usually be stopped by reducing the electrochemical potential of the metal, thereby making it cathodic and stopping anodic dissolution; a further proof of the electrochemical mechanism. In atmospheric conditions the same positive protection may often be obtained by sacrificial metal coatings. High relative humidity is usually needed for stress corrosion.

Despite the many investigations and the mass of evidence assembled on stress corrosion we are still far from a full understanding, although it is now certain that there is not a single mechanism but rather a range of mechanisms, more than one of which may be operating at any one time. Progess has been made in establishing the main parameters and in advancing credible models, but there remains an almost complete lack of knowledge on the dynamic conditions at the propagating tip of the crack where bonds are stretching and breaking, and where new metal surfaces form and react with a very small volume of solution which must in consequence develop a composition very different from the bulk of the solution. The movement of dislocations towards the base of the crack and hydrogen diffusing into the metal are also probable factors which must be fitted into concepts of anodic and cathodic reactions and the passage of ionic currents between them. Conditions cannot easily be investigated directly at the crack tip and indirect evidence of the reaction processes is sparse. It is probable that a complete explanation of stress corrosion mechanism(s) will be slow in coming.

Conditions for stress corrosion have been summarised by Fontana (1967) as:

> $(\text{susceptibility})^n \times (\text{environmental aggressiveness})^m \times (\text{stress})^p$ is a constant for the system.

If one of these factors is zero, cracking will not occur. If one factor is low then one or more of the others will need to be relatively high for cracking to occur at the same rate.

Methods of combatting stress corrosion aim at reducing one of these factors to zero at possible sites of initiation. They include:

- — modification to the alloy composition, e.g. reducing carbon to very low levels in stainless steels, or adding stabilising elements which either prevent minor phases from forming, or modify their structure if they do form, e.g. niobium or titanium in austenitic stainless steel, or silver or chromium in high strength aluminium/magnesium/zinc alloys;
- — variations in heat treatments, e.g. step quenching or overageing with high strength aluminium alloys;
- — additional low temperature heat treatments to relieve internal stresses;
- — ensuring a smooth surface finish with rounded edges to avoid sites where corrosion may readily start or where stress may be concentrated;
- — design and assembly processes which avoid high stresses from causes such as tight fits, bolted assemblies, ill-fitting or overtightened threads (or better by avoiding threaded components);

- control of welding or brazing procedures; sometimes a full post-welding heat treatment of the whole assembly is needed;
- introduction of compressive stresses into the surface layers through shot blasting or shot peening;
- the use of sacrificial coatings, e.g. zinc on aluminium, or copper alloys, or cadmium on high strength steels, to give cathodic protection (the cadmium is best deposited by a non-aqueous process, e.g. vacuum deposition, since hydrogen introduced during electrodeposition cannot be reliably removed by baking as the cadmium coating is impervious to hydrogen; the steel may then fail through hydrogen embrittlement; see later in this chapter);
- in solution, applied potential to give cathodic or, sometimes, anodic protection;
- inhibitors in solutions or as ingredients of paints or jointing compounds (care is needed since anodic inhibitors sometimes increase the risk);
- avoiding aggressive environments, especially when they contain particular contaminants, e.g. chlorides for austenitic stainless steels, especially above about 100°C; hot nitrates and hydroxyl ion concentrations for mild steel; and ammonia or amines with copper alloys. However continuing discoveries of new environments which cause stress corrosion have led to decreased emphasis on the specific role of the composition of the contaminant;
- control of working stresses at levels below the stress corrosion threshold. There is still room for debate on the existence of a general threshold, but tests on specimens which have been pre-cracked by prior fatigue in corrosive environments by the method pioneered by Brown (1968) are of value in setting thresholds for a given combination of alloy, contaminant and temperature.

A selection of the many publications on stress corrosion is given at the end of this chapter. The review of Professor Parkins (1972) is worth early reading for his overview of stress corrosion as a mixture of processes within a spectrum of mechanisms, from where stress is the major factor and corrosion assists brittle fracture through a range where electrochemical factors become of increasing importance, to the other extreme of stress-assisted intergranular corrosion, where corrosion is the dominant factor. This balanced overview makes it easier to reconcile the various detailed mechanisms proposed in the accounts of other workers.

A variety of methods is available for testing for susceptibility to stress corrosion (see Parkins, Mazza, Royvela & Scully (1972)). Slow tensile tests on pre-cracked specimens contained in a suspect environment are useful in establishing the environments which give rise to stress corrosion for a given alloy. Testing should be carried out over a range of temperatures since many alloys display temperature thresholds above which susceptibility increases rapidly.

Rolled or extruded alloys have an elongated grain structure and susceptibility to stress corrosion varies with the direction of stress in relation to the grain. The long axis of the grain, the direction of rolling or extrusion, shows least susceptibility; the greatest susceptibility usually occurs along the short transverse axis. A wide variation in susceptibility with grain orientation is frequently encountered in the high strength aluminium alloys. This has led to many modifications of processing, and variations in minor alloying additions to obtain improvements in grain structure. Rapid solidification processing (RSP) for instance, is reported by Driver (1985) to give very fine grained Al/Zn/Mg/Cu alloys which are superior to previously employed Al/Zn/Mg conventionally processed alloys which were prone to stress corrosion. This deleterious characteristic is further exaggerated in the conventional alloy when copper is added to increase the strength. Driver (1985), and Southgate (1985) also describe how grain boundaries have been eliminated in a range of nickel alloys by directional solidification (DS) to give single crystal structures; although resistance to stress corrosion is not specifically claimed, the process has obvious promise in countering susceptibility to stress corrosion.

CORROSION FATIGUE

Metals subject to cyclic stressing in an environment which is capable of reacting with the metal, may suffer cracking and failure sooner than they would under the same cycle of stresses in a non-reactive environment. This form of accelerated fatigue failure under the joint action of cyclic stress and corrosion is known as corrosion fatigue. The environment which stimulates corrosion fatigue may not be corrosive in the absence of cyclic stress since the reactions may be with the base metal exposed by a stress induced crack in the oxide film. Alternatively adsorption on the new surfaces forming behind a propagating crack reduces the energy required for crack growth and hence accelerate cracking.

In atmospheric conditions oxygen and water are the main reactants but other contaminants, notably traces of salt and acids, may exacerbate the effect. For some metals such as aluminium and steel, water vapour is usually associated with early failures and relative humidity is therefore important. For copper and lead the partial pressure of oxygen in the surrounding atmosphere is the major factor and humidity has little effect.

Corrosion fatigue reduces the life, or number of stress cycles to failure, at a given level of applied stress. More serious is the absence of a clear fatigue limit (the level of cyclic loading below which a component has an indefinite fatigue life). Under conditions of corrosion fatigue, failure is liable to occur ultimately at very low loadings, which presents engineers with a difficult design predicament.

Corrosion fatigue differs from stress corrosion in that it is seldom related to a particular combination of alloy and contaminant but occurs at varying levels of severity in most environments. Where corrosion is a major factor, in immersed or in wet contaminated atmospheres, the cracks are often branched in contrast to the single crack typical of dry fatigue.

The mechanism of corrosion fatigue is more difficult to elucidate than that of stress corrosion and has not been the subject of as many investigations. Studies indicate that the mechanism involves fracture of the oxide film under the cyclic stresses, reaction of the environment with the exposed metal, migration of dislocations and vacancies to the tip of the advancing crack, and processes of absorption, welding and tearing apart of the sides of growing cracks as the stresses cycle from compression to tension. With some metals, particularly the high strength steels, absorption of hydrogen is also involved. Useful discussions of the mechanism and occurrence of corrosion fatigue are to be found in the seminar proceedings edited by Parkins & Kolotyrkin (1980).

Susceptibility to corrosion fatigue can be reduced: by introducing compressive stresses into the surface layers of the metal by grit blasting, or shot peening, or for steel by nitriding or carburising; by anodic metal coatings such as zinc or aluminium on steel; by inhibited coatings especially within crevices, although this is less effective than sacrificial coatings, and through good design practice especially by avoiding stress-raising notches, burrs or rough machining.

FRETTING CORROSION

Fretting corrosion may result when two surfaces in contact experience continuing slight periodic movement with respect to one another. The interaction between the two surfaces is known as 'fretting' and the result 'fretting corrosion' or 'fretting damage'. The same phenomenon occurs at the contact point betweeen curved surfaces or between a curved and a flat surface when they experience slight relative periodic movement; the resulting damage is then sometimes described as 'brinelling', or 'false brinelling' if the environment is corrosive. Reaction with the environment is a secondary effect with fretting corrosion; the major cause of damage is the physical effects of the relative movement under load. Contaminants or high humidities have little influence on the severity of the damage and there is no electrochemical mechanism involved.

Fretting attack starts through high loading and subsequent deformation of the asperities of the metal surface which are the actual points of contact. The concentrated pressure at these points causes them to stretch to their elastic limit, then deform plastically, yield and tear. The intense working leads to heating of these areas with consequent high temperature oxidation, and some high pressure welding across the interface which then tears. The corrosion product is a mixture of oxide and metal particles which vary in size from 0.01 μm to 0.05 μm. The free metal content in the debris varies; levels of 13% have been reported from nickel by Fink & Hofman (1932) and 23% from aluminium have been reported by Andrew, Donovan and Stringer (1968).

The damage produced by fretting varies from large shallow depressions with roughened surfaces to smaller and deeper pits. The former are associated with conditions in which the corrosion product can readily escape and the latter where it remains trapped. As the volume of the corrosion product is greater than that of the metal from which it is formed, distortion and sometimes seizure may result at hinges and joints where the fretting product accumulates. Fretting damage increases with hardness and with the smoothness of the metal.

The cyclic movements which give fretting typically arise from the vibrations induced by the presence of heavy machinery, through travel over rough surfaces on road or rail, through rough seas or between the strands of wire ropes as they pass over pulleys or are otherwise flexed. The first effect is a discolouration of the contacting surfaces; further fretting produces roughened areas on the metal surfaces and quantities of finely powdered corrosion product which, on steel, is reddish brown and sometimes called 'cocoa', and on aluminium and most other metals is black. The colour and appearance of these fretting products is in marked contrast to the brown rust on steel, or the white corrosion product on aluminium or zinc, produced during normal atmospheric corrosion.

Mechanical assemblies where fretting corrosion is likey to be encountered include bolted or rivetted joints, pivots, ball and roller races, pressed fits, couplings, leaf springs and wire ropes. The vibration induced during transport is a major cause of fretting — curiously motor cars designed to withstand the wear and stresses of transport under their own power are prone to fretting damage on bearings, gear wheels, cylinders and leaf springs when they are transported as passive loads.

Fretting of metal trays on metal shelves has been a source of problems in the delivery of certain foods. Fretting damages the trays and shelving, but worse, produces a fine black dust of fretting corrosion product which readily disperses in the air space on disturbance and settles to spoil the appearance of exposed food and packages. A similar effect investigated by Andrew, Donovan and Stringer (1968) arose in the holds of ships carrying stacked boxes of ammunition over long voyages. The boxes and supports were of aluminium, and quantities of fretting product formed between the boxes, and at points of contact between the boxes and supports. The product contained metallic aluminium and could be ignited; when the fine dust was dispersed in air an explosive mixture resulted.

Fretting is a frequent cause of damage to stacked metal sheets during their transport. A foreign body or distortion at the base of the stack can result in an imprint of fretting damage extending through the stack. This phenomenon, often called 'transit rub', has been most reported with tinned steel, aluminium and copper; instances on all three metals with a valuable account of practical ways of avoiding it are described by Scott and Skerrey and by contributors to a subsequent discussion (1970). The stains of even slight fretting are difficult to remove, and more serious fretting soon leads to irreparable damage. Tight binding and wedging of stacked sheets is sometimes effective in overcoming the problem, but these methods may increase the loading and reduce the extent of the movement without stopping it entirely, and since high loading and a very small degree of movement cause rapid fretting, damage sometimes increases.

The relative movement, or slip, between two fretting surfaces is usually within the range 20–200 μm; wear is proportional to the number of cycles and it can become very much greater if an intermittent load is being applied normal to the fretting surface. The extent of damage is not simply related to the slip distance but approaches a maximum at a low level of slip.

Fretting damage may be prevented by separating the surfaces, by interleaving sheets of wrapping paper, for instance, between stacked sheets. Soft metal coatings such as cadmium on steel greatly reduce the damage as do many conversion coatings such as phosphating or suphurizing of steel, and anodizing of aluminium. Coatings of oil grease and temporary protectives provide protection although with the lightest coatings protection may be short lived; best effects result from the thicker and stronger resin-bonded or bitumenised products.

A good account of all aspects of fretting corrosion is given by Waterhouse (1972).

FRETTING FATIGUE

The fatigue life of steels may be seriously reduced by fretting; the combined effect is known as 'fretting fatigue'. The oscillating movement of fretting and the cycling stress of fatigue are often associated and may be occasioned by the same source. The fine pitting caused by fretting acts to concentrate stress and small cracks often associated with the fretting process may continue to grow through fatigue. Typical fretting fatigue failures start within the area of fretting damage as a fine network of cracks which spread into the surrounding metal. In tests on steel specimens fretting has been shown by Field & Waters (1967) to reduce the fatigue limit by as much as 77%. Hines (1969) reported that titanium surfaces in contact are particularly susceptible to fretting fatigue.

FILIFORM CORROSION

Filiform corrosion develops its characteristic thread-like pattern on metal surfaces under lacquer, paints or paper wraps. The growing heads of the threads spread in a random manner from a point of damage to the film where the metal surface has been contaminated. The growing heads do not cross existing tails but turn away before contacting them. Typical examples of filiform corrosion on steel and aluminium are shown in Fig. 5.7(a) and (b).

The filiform threads on steel are 0.1–0.5 mm thick. They grow at a rate of about 3 mm a week at relative humidities above about 65%; the precise level of humidity depends on the metal and contaminant combination. As the humidity increases, the thickness of the growing thread increases. On steel the threads are brown with the head showing a greenish hue.

The mechanism of filiform corrosion of steel is illustrated in Fig. 5.8. Ferrous ions from anodic dissolution at the tip of the growing thread diffuse

Fig. 5.7a — Filiform corrosion on lacquered steel. (©Controller, Her Majesty's Stationery Office 1986.)

towards the cathode where they combine with the hydroxyl ions produced by the cathodic reaction and green ferrous hydroxide precipitates; the ferrous hydroxide absorbs further oxygen and is converted to brown hydrated ferric oxide — rust. Oxygen and water diffuse through the lacquer and are absorbed into the electrolyte where they provide the essentials for further reaction. Soluble ions in solution diffuse towards the head to maintain the ionic balance through the solution. Hydrogen, being more mobile, concentrates in the tip to give acid conditions in the region of the anode and stimulates the further dissolution of iron. Some of the hydroxyl ions formed at the cathode diffuse sideways between the lacquer and steel to reinforce the oxide film there, and render the area passive and slightly alkaline which prevents attack if another active head approaches.

The attack on the steel from filiform corrosion is slight but it is unsightly and undermines the paint or lacquer so that more serious corrosion may

Fig. 5.7b — Filiform corrosion between mating surfaces on aluminium. (©Controller, Her Majesty's Stationery Office 1986.)

begin. Heavy phosphating and etch primers on steel, and anodizing on aluminium prevent filiform corrosion.

HYDROGEN EMBRITTLEMENT

Hydrogen is frequently absorbed by metals and gives a range of failure modes. Rogers (1962) distinguished four types: hydride embrittlement, hydrogen attack and two types of 'true' hydrogen embrittlement. The first is suffered by those metals which absorb large quantities of hydrogen to give hydrides, and the last three, by metals which physically absorb small amounts of hydrogen, the so-called 'endothermic occluders'. The sources of hydrogen are legion, most commonly from reaction of the metal with water during melting, casting, hot working, welding, acid treatments and electrodeposition. The metals most frequently affected are steel, titanium, copper and silver, but vanadium, zirconium, tantalum, thorium, cerium, cobalt and nickel will also absorb hydrogen and may be embrittled by it.

Hydrogen attack is the permanent effect which results from reaction between absorbed hydrogen and minor phases. In silver and copper, water forms by reaction with the oxide phase present at the boundaries of the metal grains, and the metal loses ductility. In steel hydrogen may react with carbide within the pearlite to give methane and some decarburization with a

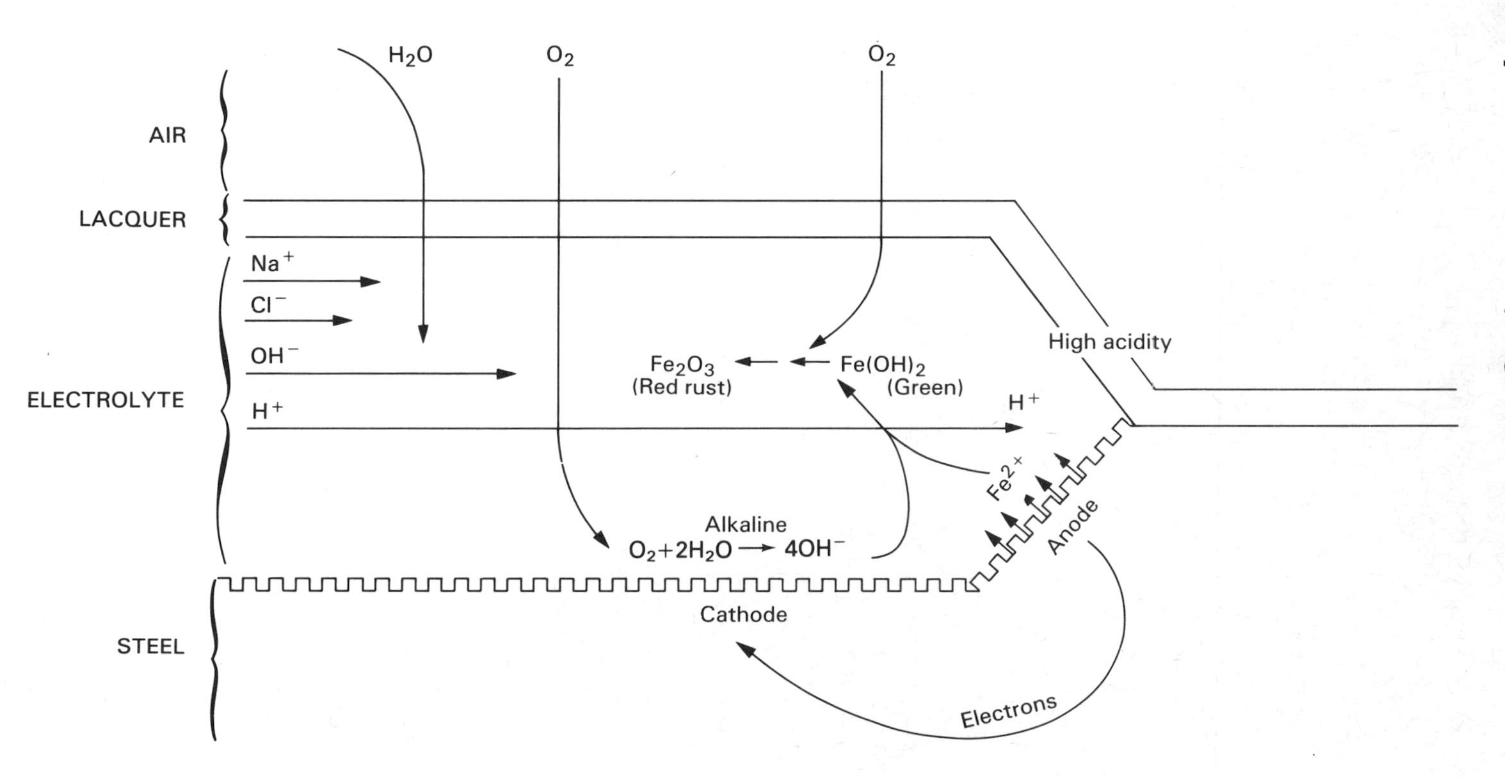

Fig. 5.8 — Mechanism of filiform corrosion.

consequent loss in ductility. Attack is prevented either by care in processing or by using resistant metals, such as oxygen-free copper or austenitic steel.

Hydride embrittlement occurs with the 'exothermic occluders' vanadium, titanium, zirconium, tantalum and thorium when hydride phases form. Typical failures were reported by Hines (1969) in describing intensive investigations into failures of titanium in chemical plant at temperatures above 100°C and with hydrogen contents of 100–200 parts per million. Hydrogen pick-up was associated with damage to the oxide film and iron contamination on the surface. Diffusion into the metal was only detectable above 100°C. The problem was overcome by anodizing which removed impurities and reinforced the oxide film.

The two forms of 'true' hydrogen embrittlement encountered are: low strain rate embrittlement or sustained load failure, and the flakes, cracks, fractures and 'fish eye' inclusions which may be encountered in ingots and welded structures. These fractures are the result of sustained load failures due to internal strain from differential heating or cooling interacting with absorbed hydrogen and are of course irreversible. The slow strain rate embrittlement of high strength steel is the form of hydrogen embrittlement most likely to be encountered in storage and it is discussed further in the following paragraphs.

High strength steels containing hydrogen may fail under a sustained stress which is well below their ultimate tensile strength. As little as 30% of the ultimate tensile stress may be sufficient to induce failure. The concentration of hydrogen involved is often well below 0.1 parts per million and the time to failure varies from minutes to months. The hydrogen may be introduced during manufacture and forming of the metal, or in subsequent finishing processes. Acid pickling, electroplating and phosphating are typical of the processes which introduce hydrogen into steel fabrications.

Susceptibility of steels to hydrogen embrittlement increases with the strength of the steel. For precautionary procedures against hydrogen embrittlement, DEF STAN 03–2 and DEF STAN 03–4 consider steels in four categories of increasing strength (or corresponding surface hardness). These limits with an indication of the severity of the hazards are given in Table 5.1.

Steels case-hardened by carburising or nitriding over any part of their surface must be considered in the group appropriate to the hardest portion of their surface, and since this is frequently above 600 HV, such steels are usually in group 4. The same precautions need to be applied to steels selectively case hardened by electron beam hardening, laser hardening, plasma nitriding and ion implantation, but there is a lack of test results or recorded experience on steels treated by these processes. Martensitic stainless steels are embrittled by hydrogen to a similar degree to low alloy steels of similar strength. Maraging steels are somewhat less susceptible although they are included in the same groupings for de-embrittling procedures. Austenitic stainless steels are unaffected.

Although hydrogen embrittlement is a major hazard with strong steels, the vast majority of steel is manufactured to strengths within group 1 and is

Table 5.1 — Groups of steels for susceptibility to hydrogen embrittlement (from DEF STAN 03–2 and DEF STAN 03–4)

Group	Max. tensile strength (N/mm^2)	Hardness (HV)	Sustained load sensitivity
1	Up to 1000	<340	Insensitive
2	1100–1450	340–440	Some sensitivity
3	1450–1800	440–560	Sensitive: a serious problem
4	Over 1800	Over 560	Very sensitive: a most serious hazard

therefore unaffected. In general engineering the most frequently encountered components susceptible to hydrogen embrittlement are springs, prestressed concrete tensioning steels, pipelines, piano wires and case-hardened steels. Springs may fall in groups 2, 3 or 4; the tensioning steel for prestressed concretes are usually in group 2; piano wires are either group 3 or 4, and case-hardened steels are frequently in group 4. Group 2 steels are used in critical applications such as guns and drive shafts. Group 3 and 4 steels are most frequently encountered in aircraft and rocket motors but even in these applications, where the savings in weight offered by increased strength can frequently justify careful manufacture and regular inspection, the high risks associated with group 4 steels often exclude them from use. Cathodic protection should not be used with steels in groups 2–4.

Two mechanisms are favoured in accounting for the phenomenon of hydrogen embrittlement. One postulates that mobile hydrogen atoms concentrate around dislocations in the body of the metal and hinder their movement and so prevent the slip and slight plastic deformation needed to distribute stress evenly through the steel. The steel then behaves as a brittle material since the mechanism responsible for its toughness is blocked. The other hypothesis considers that hydrogen collects around dislocations and is carried by them to points of high stress such as the roots of notches or the tips of cracks in the normal process of slip and dislocation movement. At the point of high stress the hydrogen weakens the cohesion of the steel and lowers the energy of free surfaces so that they form more readily. The net effect is to lower the total energy of crack propagation. There is evidence favouring both mechanisms and it seems probable that both contribute.

Hydrogen diffuses readily in steel, and atoms which reach a surface may combine and escape as hydrogen molecules. These processes of diffusion, combination and escape accelerate with temperature and are relatively fast above about 150°C. Therefore after any processing which might induce hydrogen pick-up, susceptible high strength steels should be baked above 150°C for a period sufficient to remove susceptibility, typically 1 hour. The temperature and time required depend on the strength of the steel, the type of surface coating, the process that has generated the hydrogen, and the cross-section of the component. Cathodic cleaning treatments and electrodeposition give rise to the highest levels of hydrogen and these treatments should be avoided in processing groups 3 and 4 steels since some permanent damage may be caused by the high concentration of hydrogen and residual surface stresses. It may also be too difficult to reduce the hydrogen to safe levels particularly with electrodeposition since some metals coatings are impervious to hydrogen and therefore delay or prevent its escape.

Some impurities, notably sulphur, selenium and arsenic, inhibit the second step in the release of hydrogen so that the free atoms remain longer on the surface and more of them diffuse into the metal. For this reason sulphur containing brightening agents should not be used in plating any high strength steels, and if sulphuric acid is used for cleaning, oxidisable impurities which might lead to sulphur release should be kept below 10 parts per million. For group 4 steels electrodeposited coatings are avoided and where metal coatings are needed non-aqueous treatments should be used; in aircraft construction vacuum-deposited cadmium coatings have been preferred to avoid the introduction of hydrogen.

WHISKERS

Two types of whiskers may grow on metal surfaces. The first are metallic whiskers which develop on the metals of relatively low melting point, tin, cadmium and zinc, and the second are the silver sulphide whiskers on silver.

Metal whiskers typically of diameter about 1–5 μm grow to lengths of up to 30 mm, on relatively pure metal coatings which are under stress. The stress is usually internal, from plating either at low temperatures, or from solutions containing brightening agents which often give highly stressed electrodeposits. The stress may also be transmitted from surface stresses in the substrate, or arise from impact damage, or it may be applied externally through bolted joints for instance. The growth rate of metallic whiskers varies but 3 mm a year is typical. They grow mostly as straight parallel-sided single crystals but occasionally change direction, usually at angles suggesting growth on an alternative crystallographic plane. Tin whiskers on brass substrates are shown in Fig. 5.9(a) and (b).

Whiskers appear to grow from screw dislocations. The energy for their growth is produced by the relief of stress, and the mechanism is by migration of dislocations and their relief through regular atomic migrations centred on arrays of rotating edge dislocations.

The major problems arising from whisker growth have been electrical shorting across the faces of the plates of capacitors and between conductors in micro-switches and micro-circuits. Sugiarto, Christie & Richards (1984), in investigations on components for electrical systems, report zinc whiskers on bright electrodeposits up to 30 mm in length and others capable of passing a continuous current of 5 mA with a fusing current of 7 mA. Dull zinc

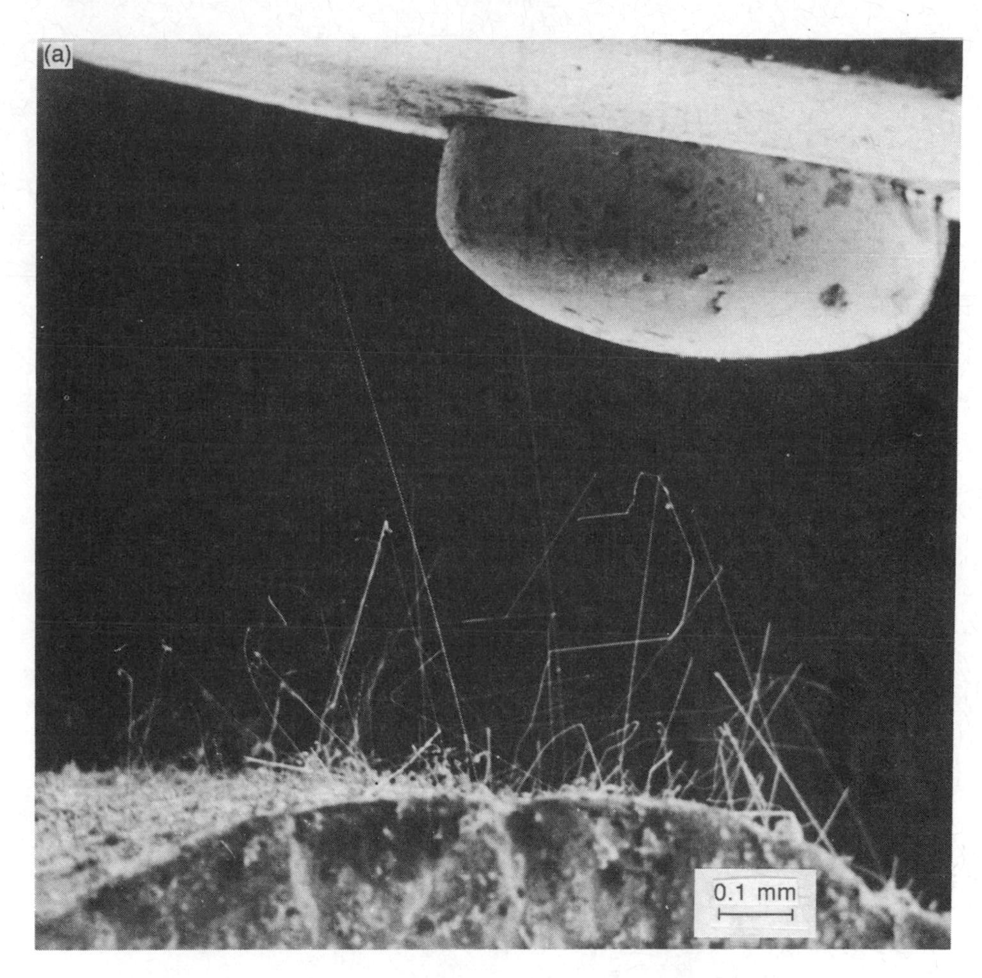

Fig. 5.9a — Tin whiskers on electrodeposited tin on brass: whiskers on lower contact of switch. (©Controller, Her Majesty's Stationery Office 1986.)

electrodeposits from cyanide baths without organic additives did not produce whiskers.

The growth of metal whiskers may be reduced by addition of 1–2% of alloying elements and by avoiding brightening agents in electroplating baths and other causes of stress, but metals of low melting points are best avoided on closely spaced conductors.

Silver sulphide whiskers grow on silver in confined atmospheres which contain high concentrations of sulphur dioxide or hydrogen sulphide. The whiskers are similar in form to metallic whiskers and will also conduct electricity so that their growth may give similar electrical problems to metallic whiskers. Fig. 5.10 shows a typical example of silver sulphide whisker growth on a silver-coated component which was enclosed with a polysulphide rubber.

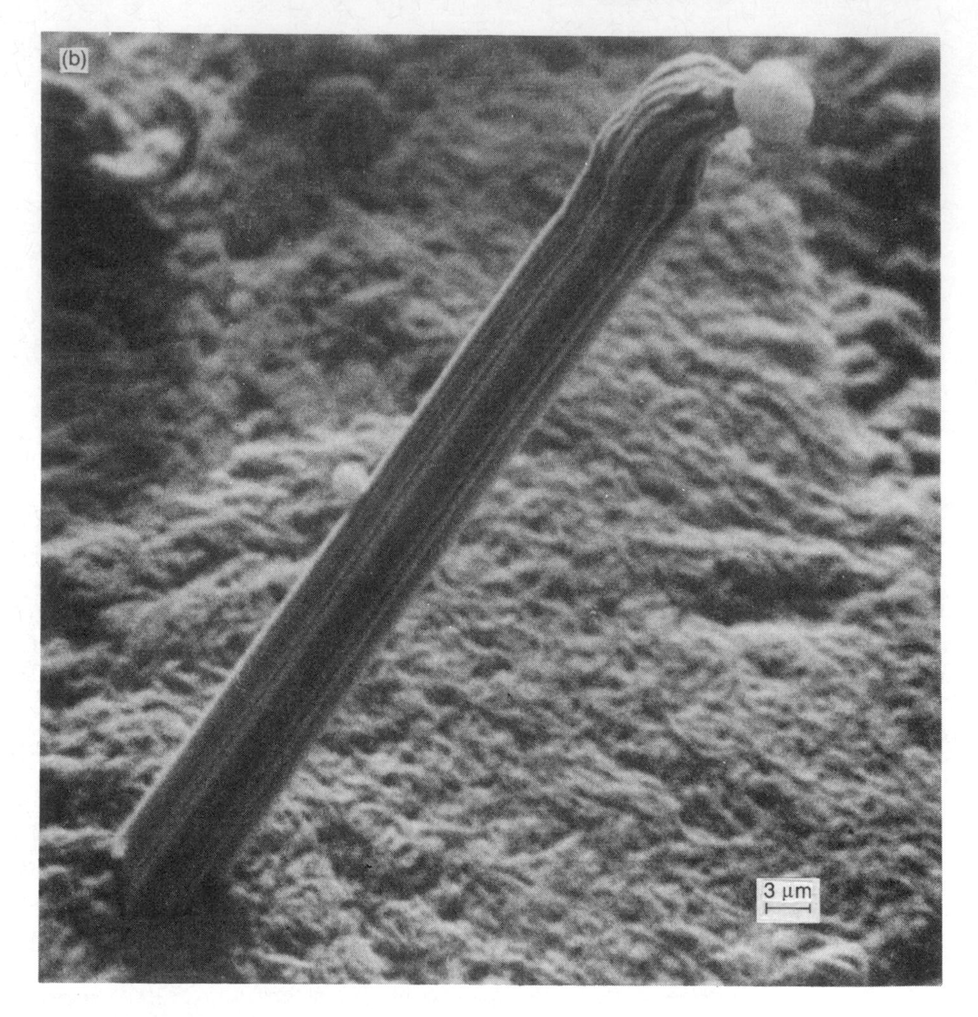

Fig. 5.9b — Tin whiskers on electrodeposited tin on brass: detail of whisker. (©Controller, Her Majesty's Stationery Office 1986.)

Tracking between silver-plated connectors on printed circuit boards also occurs through silver migration but this happens through a process of corrosion and redeposition, and growths spread laterally over the board; this type of failure is associated with high humidity whereas whiskers may grow in drier conditions.

TIN PEST

At temperatures below 20°C metallic tin is metastable; an enantiomorphic-form, grey tin, is the stable form at lower temperatures. The change to grey tin at lower temperatures is most readily started by contact with a 'seed' of grey tin and then spreads slowly from the point of contact through the

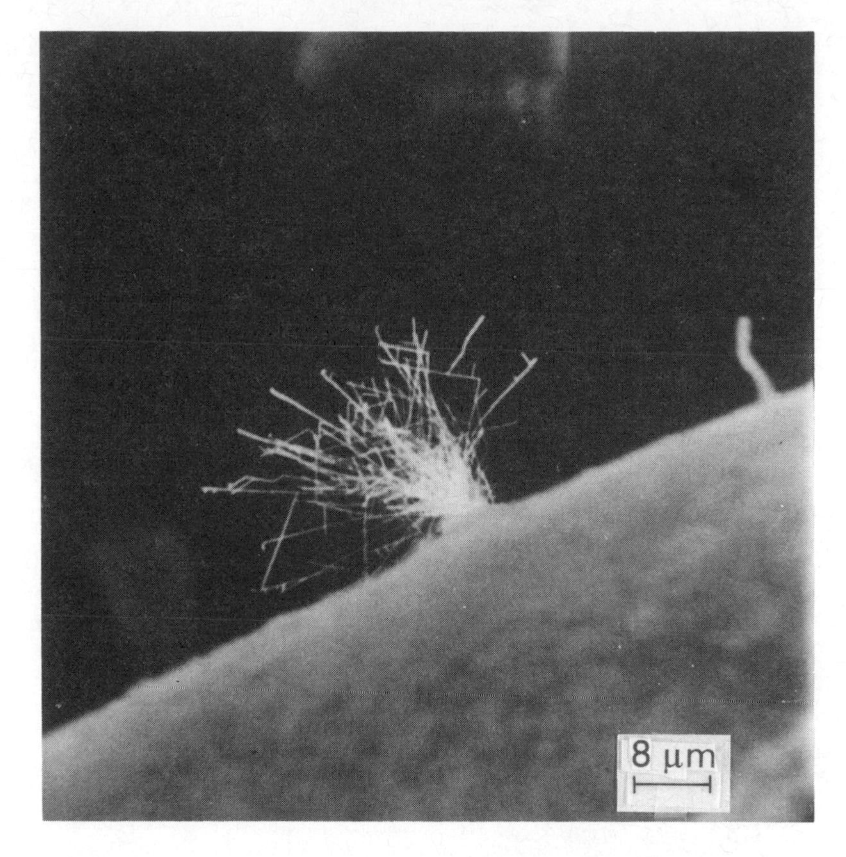

Fig. 5.10 — Whiskers of silver sulphide on electrodeposited silver: after storage enclosed with polysulphide rubber. (©Controller, Her Majesty's Stationery office 1986.)

formation of brittle material with surface blistering, the whole eventually disintergrating into a sandy powder.

This phenomenon, known as tin pest, has been encountered with tin organ pipes in cold churches, and with tin buttons, ingots and solders. Tin pest is rarely encountered nowadays since pure tin is seldom used and small amounts of alloying additions prevent the transformation.

LIQUID METAL CORROSION

Liquid metals may cause a form of stress corrosion cracking of some alloys. Mercury for example produces rapid cracking of stressed hard brass and this effect is often used as a test for susceptibility of brass to atmospheric stress corrosion cracking. Failure of stressed high strength steel and titanium alloys may occur by 'liquid' metal attack from contact with the low melting point metals: cadmium, zinc, lead and tin at temperatures above about 150°C. Failures are most frequently associated with the metals used as protective coatings which have been applied either directly to the susceptible metal or on an adjacent metal, but failure may arise from any contact. Nicholson, Towers & Partington (1983), for instance, give details of liquid metal embrittlement of steel by brass and refer to other examples such as the embrittlement of stainless steel by zinc. The liquid metal lowers the surface energy, change for grain boundary cracking through alloying with the new surfaces at the crack tip. Failures can be caused by sustained stress (stress corrosion cracking) or by cyclic stressing (corrosion fatigue).

Mercury has a damaging effect on some metals through the amalgam it forms. On aluminium it is especially damaging since the aluminium oxidises at the surface of the amalgam away from the solid surface of the aluminium. The oxide grows as fungus-like excrescences and the mercury is released to eat further into the metal. The fear of mercury attack on the internal surfaces of the aluminium skins of aircraft has led to very stringent regulations for the carriage of mercury by air. Once mercury has escaped it is very difficult, or more correctly nearly impossible, to decontaminate the aircraft. Beard & Hine (1965) studied the effect of alloying constituents in aluminium on the corrosive attack by mercury. They showed that additions of copper, silicon and manganese inhibited attack; zinc, magnesium, chromium, iron and titanium did not affect attack, which was as devastating on alloys with these additions as it was on unalloyed aluminium.

SELECTIVE CORROSION (DE-ALLOYING)

Corrosion of some alloys takes place through dissolution of the more reactive components, with the visible effect limited to a change of colour or appearance, while the selective dissolution eats deeply leaving a porous deposit having little strength or cohesion.

The most widely encountered example is the dezincification of brass. The zinc of the alloy dissolves and a spongy residue of copper remains. At the sites of attack the brass assumes a coppery appearance and may be marked by surrounding stains. What appears to be a superficial discoloration is in practice the site of penetrating attack into the alloy. It is not clear whether the more reactive zinc is selectively dissolved or if initially both elements are dissolved and the less strongly electropositive copper ions are redeposited by displacing zinc from the surface. Cast iron also suffers selective dissolution; the iron dissolves leaving a skeleton of graphite. This effect is frequently encountered in immersed and buried structures but seldom in atmospheric corrosion since the bulky rust in part protects the metal from attack and also stresses and disrupts the graphite skeleton as it forms.

REFERENCES

Andrew, J. F., Donovan, P. D. & Stringer, J. (1968) *Br. Corros. J.* **3,** 86.
Beard, F. M. & Hine, R. A. (1965) *Br. Corros. J.* **1,** 98.
Brown, B. F. (1968) *Metallurgical Reviews* **13,** 171.
Driver, D. (1985) *Metals & Materials* **1,** 6, 345.

Evans, U. R. (1963) *The Corrosion and Oxidation of Metals*. Edward Arnold.
Field, J. E. & Waters, D. (1967) *NEL Report,* No. 275.
Fink, M. & Hofman, U. (1932) *Metallkunde* **24,** 49.
Fontana, M. G. (1967) *Proc. AGARD Conf.* 1–3.
Hines, J. (1969) *Br. Corros. J.* **4,** 4.
Johnson, K. E. & Abbott, J. S. (1974) *Br. Corros. J.* **9,** 3, 171.
Mazurkiewicz, B. (1983) *Corros. Sci.* **23,** 7, 687.
Mazurkiewicz, B. & Piotrowski, A. (1983) *Corros. Sci.* **23,** 7, 697.
Nicholson, C. E., Towers, R. T. & Partington, V. (1983) *Metallurgist & Mat. Techn.* **15,** 6, 285.
Parkins, R. N. (1972) *Br. Corros. J.* **7,** 15.
Parkins, R. N., Mazza, F., Royvela, J. J. & Scully, J. C. (1972) *Br. Corros. J.* **7,** 4, 154.
Parkins, R. N. & Kolotyrkin, Ya. M. (Ed.) (1980) *Corrosion Fatigue*. Proc. USSR–UK Seminar on Corrosion Fatigue, 19–22 May 1980).
Robinson, J. J. (1983) *Corros. Sci.* **23,** 8, 887.
Rogers, H. C. (1962) *Materials Protection.* April 26.
Scott, D. J. & Skerrey, E. W. (1970) *Proc. Conf. Protection of Metal in Storage and Transit.* London, Brintex Exhibitions, pp. 45 and 54 (discussion).
Southgate, R. J. (1985) *Metals & Materials* **1,** 10, 602.
Sugiarto, H., Christie, I. R. & Richards, B. P. (1984) *J. Inst. Met. Fin.* **62,** 3, 92.
Waterhouse, R. B. (1972) *Fretting Corrosion.* Pergamon Press.

FURTHER READING

Logan, H. L. (1960) *The Stress Corrosion of Metals.* John Wiley.
AGARD (1970) *Engineering Practice to Avoid Stress Corrosion.* Conference Proceedings No. 53, NATO.
Department of Industry (1980) *Stress Corrosion. Guides to Practice in Corrosion Control,* No. 4. HMSO.
Corrosion Control, No. 14. HMSO.
Corrosion Source Book (1984) ASM & NACE.

6

Vapour Corrosion of Engineering Metals

Vapour corrosion is the acceleration of atmospheric corrosion by volatile contaminants, often present in only trace amounts. Typically these contaminants are short-chain organic acids, especially formic and acetic acids, but corrosion by other volatiles including hydrogen chloride, sulphur dioxide, hydrogen sulphide, hydrazoic acid, phenol, phosphine, nitric acid and ammonia has also been observed in practice and each of these has been recorded as causing serious corrosion of packaged goods.

The corrosive effect of acetic acid has been known since at least 450 BC when Theophrastos described a process for producing white lead from metallic lead sheet using vapours from vinegar to corrode the lead. This reaction, now known as the Dutch Stack Process, has been used to modern times. The overall reaction is:

$$2Pb+4CH_3.CO_2H \rightarrow 2Pb(CH_3CO_2)_2 \text{ (lead acetate)}+2H_2O$$

$$2Pb(CH_3CO_2)_2+CO_2+3H_2O \rightarrow Pb(OH)_2.PbCO_3\text{(insoluble)}+4CH_3CO_2H$$

The cathodic and anodic reactions at the metal surfaces are:

$$\text{anode } Pb \rightarrow Pb^{2+}+2e$$

$$\text{cathode } O_2+2H_2O+4e \rightarrow 4OH^-$$

The lead ions from the acidic area of the anode diffuse towards the more alkaline area of the cathode and carbon dioxide is absorbed from the atmosphere:

$$2Pb^{2+}+4OH^-+CO_2 \rightarrow Pb(OH)_2.PbCO_3 \text{ (white lead)}+H_2O$$

The white lead precipitates and the acetic acid is regenerated to cause further attack.

More recently, Watson (1789) commented on the profuse corrosion of the underside of the lead on certain church roofs in the vicinity of oak rafters. This he attributed to volatile acids and he observed that the corrosion was considerably less with deal. Turpentine and linseed oil were also identified as sources of corrosive vapours during the early years of this century. With greater industrialisation and the manufacture, transport and storage of increasingly complex assemblies such as radio sets, motor cars, clocks, watches and cameras, where both metals and organic materials were stored together in enclosed packages, the phenomenon became widespread.

Investigations have extended our awareness of the wide range of sources of corrosive vapours and it has now become an essential requirement to evaluate organic materials before using them as packaging for metals, or enclosing them with metals. With the explosion in the use of plastics and the increasing range of formulations becoming available, the need for continuing investigations remains.

The available information on the sources of these corrosive vapours is recounted in later sections of this chapter but in anticipation it may be said that the most frequently encountered corrosive vapour is acetic acid, and wood is the major source. Formic acid vapours are also encountered from paints, adhesives and plastics. A typical example of a vapour corrosion failure is shown in Fig. 6.1: the corroded cadmium-plated rocket fin was one of a batch which was stored in plywood boxes for six months in an unheated building in the United Kingdom. The cadmium plating which was 12 μm thick, and covered with a coating of lanolin, corroded through and attack began on the base steel. This thickness of cadmium would protect steel for over five years in a marine atmosphere.

SUSCEPTIBILITY OF METALS TO VAPOUR CORROSION

The metals most rapidly attacked by volatile organic acids are steel, zinc, cadmium, lead and magnesium, but many other metals are also affected.

A test bottle used by Donovan & Stringer (1971) for evaluating both the rate of attack of volatile acids on metals and the corrosivity of possible sources of corrosive vapours is shown in Fig. 6.2: 50 mm × 25 mm metal coupons and a suspected source of corrosive vapours are suspended from glass hooks at the base of the stopper; 2 ml of water are added to maintain the relative humidity at 100%; the capillary 1 mm diameter × 25 mm long on the stopper allows oxygen consumed in the corrosion process to be replenished; for the standard test the vessel is maintained at 30°C for 21 days when the extent of attack on the metal is determined.

The test pieces shown in Fig. 6.2 are of wood, air-drying paint, polyacetal and polyester, all of which are sources of corrosive vapours. The same method has been used to determine the attack on metals of acid vapours. Known concentrations of acetic and formic acid vapours are obtained in the

Fig. 6.1 — Cadmium-plated rocket fin (right) corroded by six months of storage in a plywood box in an unheated building. (© Controller, Her Majesty's Stationery Office 1986.)

air space in equilibrium with aqueous solutions of the acids, at appropriate strengths, determined from the relationship in Fig. 6.3

Fig. 6.4, from the work of Clarke & Longhurst (1961), shows the effect of increasing concentrations of acetic acid vapour on steel, zinc, cadmium, copper and brass. Steel and zinc are seen to be rapidly attacked in atmospheres containing even less than 1 part per million (ppm) of acetic acid.

Results of tests on a range of pure metals in three concentrations of both acetic and formic acids (from Donovan & Stringer (1971)) are given in Table 6.1.

From Table 6.1 it can be seen that zinc, lead, magnesium, iron, cadmium and manganese are rapidly attacked by both acetic and formic acid vapours, with acetic acid giving a higher rate of corrosion than formic acid. The severity of corrosion is evident from the attack of zinc, which amounts to 16 μm in the lowest concentration of acetic acid vapour (0.5 ppm). This thickness of zinc is sufficient to protect steel for two years in a marine atmosphere and considerably longer on urban exposure. Copper is attacked by both acetic and formic acid vapours but at a much lower rate. Aluminium is corroded but even more slowly. The attack on nickel is selective and pitting results but chromium is not corroded by either acid vapour, so that nickel with the traditional flash of chromium resists attack. Tin is generally resistant although some attack is encountered at high concentrations of acetic acid. Silver is unaffected by formic or acetic acids. Indium is unusually attacked more rapidly by formic acid than by acetic.

The maximum in the corrosion of zinc shown in Fig. 6.4, in atmospheres

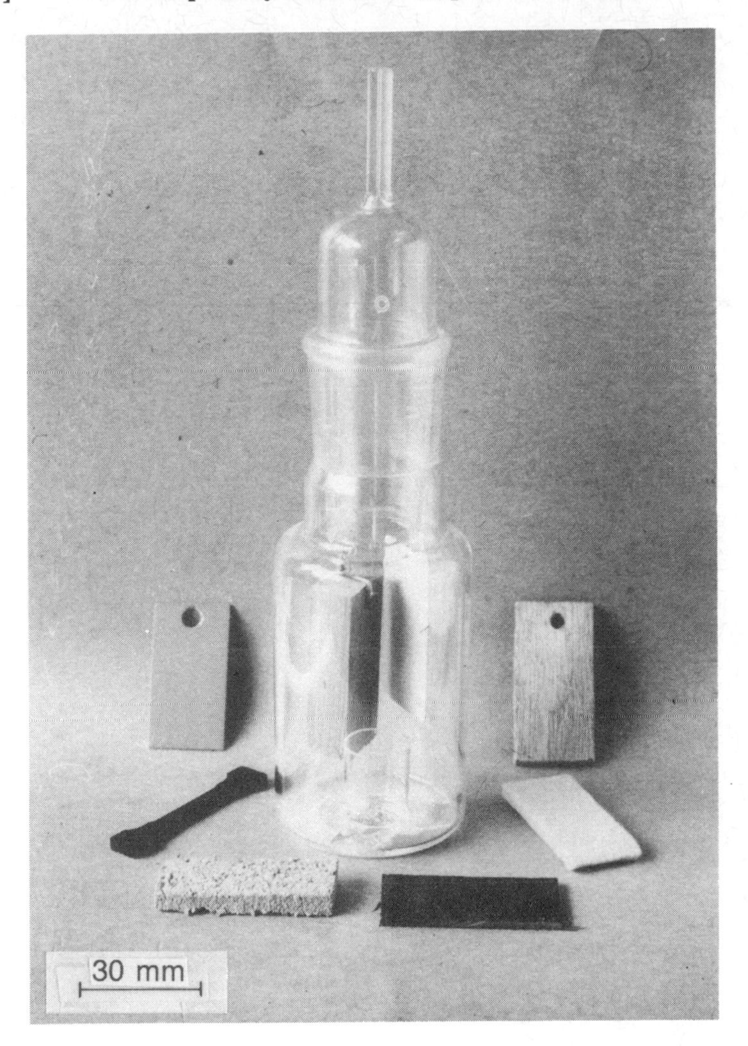

Fig. 6.2 — Vessel and test pieces for vapour corrosion evaluation. (© Controller, Her Majesty's Stationery Office 1986.)

containing 5 ppm acetic acid, corresponds with a change in the corrosion product from a loose powder of basic zinc carbonate to an adherent glass-like coating of zinc acetate. This is consistent with the reaction proposed earlier — the higher concentrations of acid do not allow hydrolysis and precipitation of carbonate.

Table 6.2, from the work of Donovan & Moynehan (1965), shows the extent of corrosion of zinc, cadmium and steel in atmospheres containing formaldehyde, and propionic and butyric acids, which are three volatiles evolved from freshly applied air-drying paints.

The rapid attack of cadmium in atmospheres containing propionic and

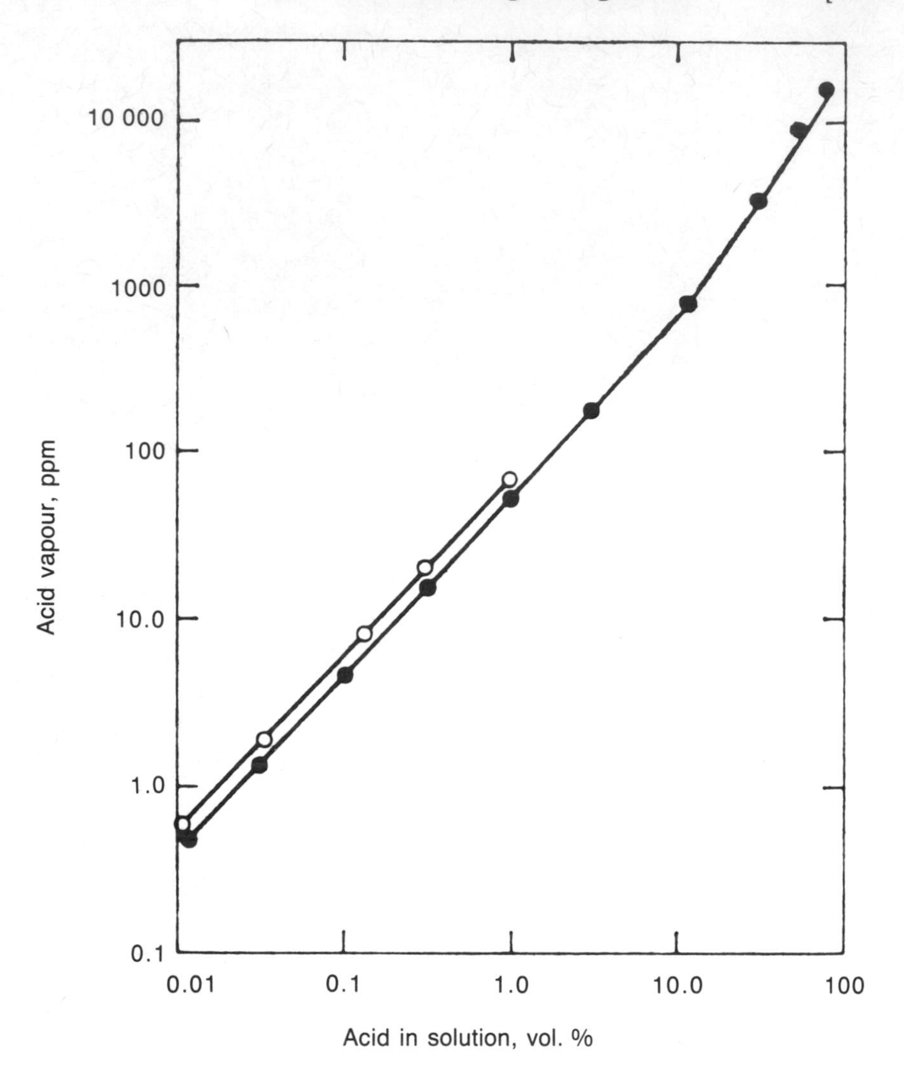

Fig. 6.3 — Relation between acetic and formic acid concentrations in equilibrium in air and aqueous solutions at 30°C. ○ formic acid; ● acetic acid. (© Controller, Her Majesty's Stationery Office 1986.)

butyric acids explains the ready susceptibility of cadmium to corrosion by vapour from freshly applied drying-oil paints. The corrosion of zinc by formaldehyde suggests that zinc is an effective catalytic surface for oxidation of formaldehyde to formic acid.

RELATIVE SUSCEPTIBILITIES OF METALS TO ORGANIC VAPOUR CORROSION

The engineering metals divide into four groups on the basis of their susceptibility to corrosion by organic acid vapours:

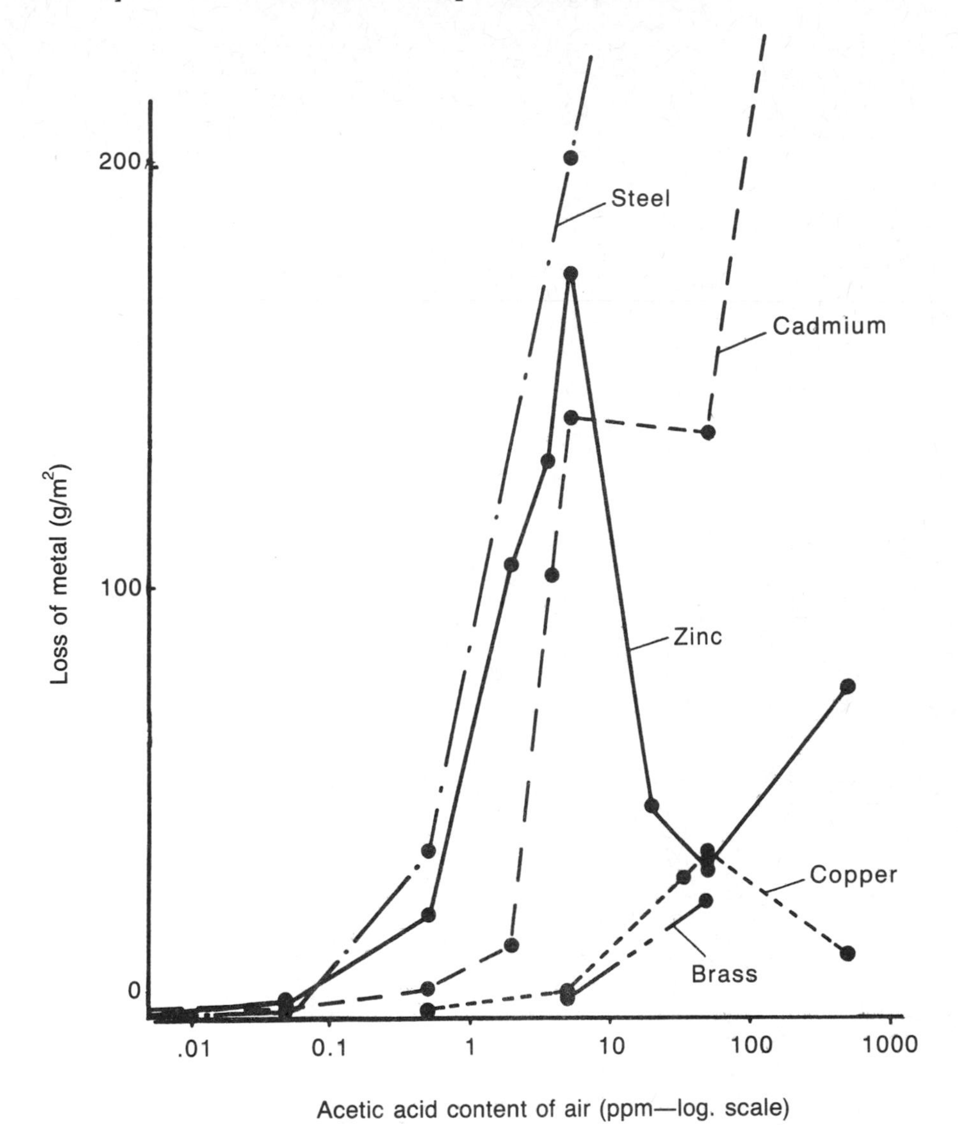

Fig. 6.4 — Corrosion of metals by acetic acid vapour: 3 weeks, 30°C, and 100% RH. (© Controller, Her Majesty's Stationery Office 1986).

— rapid attack Zn, Cd, Fe, Mg, Pb, In and Mn;
— moderate attack Al and Cu;
— slight attack Sn, Ni and Co;
— immune Ag, Ti, V, Mo, W, Cr, stainless steels and noble metals.

Detailed consideration must be given to the compatibility of all materials which are to be enclosed with any of the first group if they are unprotected. Only with the last group is there no corrosion problem with organic acids,

Table 6.1 — Corrosion attack (μm) by acetic and formic acid vapours (100% RH, 30°C, for 3 weeks) (reproduced with permission from Donovan & Stringer (1971))

	Acetic acid			Formic acid			
Metal	0.5 ppm[a]	5 ppm	50 ppm	0.6 ppm	6 ppm	60 ppm	(no acid)
Magnesium	22	130	330	5	5	18	6
Aluminium	0.4	0.7	4	0.7	1	6	0.7
Titanium	<0.02	<0.02	<0.02	<0.02	<0.02	<0.02	<0.02
Manganese		11	116		4	10	0.6
Vanadium	0.6	0.7	0.3	0.6	0.4	0.2	0.7
Zinc	16	52	10	4	8	10	0.1
Iron	3	37	44	4	13	24	0.1
Cadmium	1.1	32	39	1	2	13	0.1
Cobalt	0.2	2	6	0.1	0.7	3	0.02
Indium	0.9	0.3	12	1	10	3	0.05
Nickel	<0.01	<0.01	0.7	0.01	2	3	<0.1
Molybdenum	0.03	0.03	0.03	0.02	0.05	0.2	0.03
Tin	<0.01	<0.01	0.3	<0.01	<0.01	<0.01	<0.01
Lead	9	14	27	1	6	3	0.5
Tungsten	0.3	0.3	0.4	0.3	0.3	0.3	0.2
Copper	0.2	2	7	2	2	8	0.03
Silver	<0.01	<0.01	<0.01	<0.01	<0.01	<0.01	<0.01

a The acid vapour concentration in parts per million (v/v) in air.

Table 6.2 — Corrosion attack (μm) in air by solutions of 0.01 and 0.10% (v/v) formaldehyde, propionic acid and butyric acids (100% RH, 30°C, for 3 weeks) (from Donovan & Moynehan, 1965). (© Controller, Her Majesty's Stationery Office 1986)

	Formaldehyde		Propionic acid		Butyric acid		(No acid)
Metal	0.01	0.10	0.01	0.10	0.01	0.10	
Zinc	0.7	49	0.7	1.4	0.7	0.6	0.1
Cadmium	0.9	0.7	0.8	13	0.9	2	0.1
Mild steel	0.1	0.02	4	41	2	20	0.1

but even here organic vapours may condense and polymerise to give conducting films which may lead to difficulties with electrical contacts, and inorganic acid vapours may give corrosion. Some guidance is given on these problems in the Ministry of Defence *Guide for the Prevention of Corrosion of Metals caused by Vapours from Organic Materials* DEF STAN 03-13 (1977), and in the British Standards and Ministry of Defence *Standards and Codes on Packaging*: BS 1133, BS 4672, DEF-1234A, and DG-8.

VAPOUR CORROSION OF METAL COATINGS

The two most widely used coatings for the protection of steel in full outdoor exposure have been zinc and cadmium, although recently the popularity of cadmium has declined because of fears of its toxicity. Unfortunately these two metals are rapidly corroded in the presence of organic acid vapours and alternatives with greater resistance are often needed.

The good performace of tin plating has made this a preferred metal coating for steel if vapour corrosion is an expected hazard; coatings of either tin direct onto steel or a duplex coating of tin on cadmium have been used. Fig. 6.5 shows the results of tests in atmospheres of acetic and formic acid vapours on a series of tin and cadmium electrodeposited coatings on steel.

Electrodeposited tin gives better protection to the steel in atmospheres containing the organic acid vapours than does cadmium, although on outdoor exposure cadmium is greatly superior (see Fig. 12.1). It is interesting to observe that formic acid vapour is more aggressive than acetic acid vapour to the tin-plated steel, although as can be seen from Table 6.1, tin itself is more susceptible to attack by acetic acid. Formic acid being the stronger acid gives a solution of higher conductivity which enhances the bimetallic effects associated with attack at pores in cathodic coatings, but other factors including pH levels and the stability of the metal acid complexes are probably also involved.

The tin/cadmium duplex deposit is effective in protecting steel on outdoor exposure although its performance falls short of cadmium without the tin. Against vapour corrosion the performance of duplex Sn/Cd is dependent on low porosity in the tin coating which requires careful process-

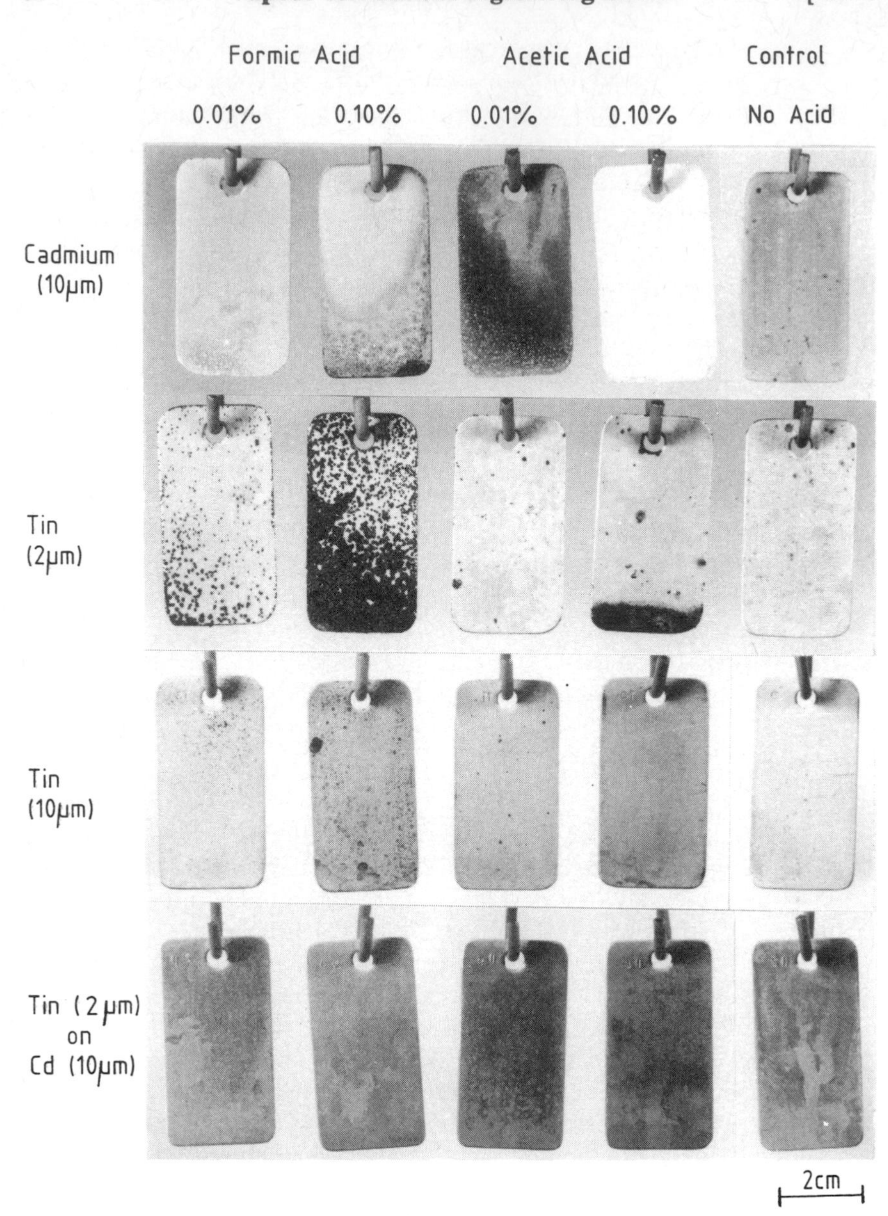

Fig. 6.5 — Vapour corrosion tests on Cd, Sn and Sn on Cd, electrodeposits on steel: 3 weeks, 30°C, 100% RH. (© Controller, Her Majesty's Stationery Office 1986.)

ing with little delay between the two plating baths. Unfortunately there are no current specifications governing the duplex coating. A range of thicknesses of both cadmium and tin has been examined by Donovan & Stringer (unpublished). The best performance was obtained with 2 μm of tin over 10 μm of cadmium, shown in Fig. 6.5. Thicker overcoatings of tin blistered on exposure; thinner coatings were porous.

A number of cadmium, zinc and manganese alloy electrodeposits have been considered by Donovan & Stringer (1971) for their resistance to attack by acetic acid vapour. The results of tests on some of these coatings together with corresponding tests on electrodeposited zinc, cadmium and manganese are given in Table 6.3

Table 6.3 — Attack on electrodeposited alloy coatings (10 μm) on steel by acetic acid vapour (100% RH, 30°C, for 3 weeks) (reproduced with permission from Donovan & Stringer (1971))

Coating	Concn of acetic acid in vapour (ppm)									
	Nil	0.005	0.05	0.5	2.0	3.5	5.0	20	35	50
Mn	0.4	0.4	0.5	7	*	*	*	*	*	*
Mn/Se (99:1)	0.5	0.8	0.7	0.5	2	2	4	*	*	*
Mn/Zn (35:65)	1.7	2.2	2.3	7	9	*	*	*	*	*
Zn	0.1	0.3	0.4	4	*	*	*	*	*	*
Zn/Fe (60:40)	0.4	0.4	0.5	7	*	*	*	7	*	*
Zn/Ni (84:16)	0.07	0.1	0.1	0.8	1	3	3	5	7	9
Zn/In (76:24)	0.4	0.6	0.8	0.7	*	*	*	*	*	*
Cd	0.3	0.3	0.5	1	*	*	*	*	*	*
Cd/In (48:52)	0.2	0.2	0.2	1	4	*	*	*	*	*
Cd/Sn (70:30)	0.1	0.1	0.2	0.5	2	4	7	8	8	9

*Extensive rusting of steel base.

All the coatings, in the tests recorded in Table 6.3, protected sacrificially, corrosion of the base steel only starting when the coatings were almost completely corroded away. Zinc/nickel, manganese/selenium and cadmium/tin were much more resistant to attack by acetic and formic acids than were zinc or cadmium coatings; all three coatings gave good protection when exposed in urban or marine environments.

Duplex nickel/chromium coatings are widely used on both steel and brass for their decorative appearance coupled with moderate resistance to corrosion in outdoor and humid atmospheres. When nickel/chromium electrodeposits on steel are exposed to atmospheres containing acetic or formic acid there is little or no corrosion of the plating but the basis metal may be attacked at pores to give the typical rust spotting. Results of corrosion tests are given in Table 6.4.

EFFECT OF RELATIVE HUMIDITY ON VAPOUR CORROSION

Fig. 6.6, from the work of Clarke & Longhurst (1961), shows the effect of a concentration of about 50 parts per million of acetic acid vapour on four metals at a range of relative humidities; the familiar dependency of corro-

Table 6.4 — Corrosion of nickel/chromium duplex coatings on steel and brass by acetic and formic acid vapours (100% RH, 30°C, for 3 weeks) (reproduced with permission from Donovan & Stringer (1971))

	Attack on plated steel and brass specimens (g/dm²)			
	[a] Nickel/chromium duplex coating on steel		[a] Nickel/chromium duplex coating on brass	
Coating thickness (μm)	10	20	7.5	15
Acetic acid solution				
0.01 vol.% (0.5 ppm in air)	0.01	0.001	0.0007	—
0.1 vol.% (5.0 ppm in air)	0.11	0.02	0.002	0.0005
Formic acid solution				
0.01 vol.% (0.6 ppm in air)	0.02	0.005	0.001	0.0007
0.10 vol.% (6.0 ppm in air)	0.28	0.06	0.004	0.002

a Plated to BS 1224:1965

sion on high relative humidity can be seen, with the corrosion rate of all four metals becoming low at 72% RH.

SURFACE TREATMENTS AND PAINT ON METALS: WITH ACID VAPOURS

Paint on zinc or cadmium often fails when enclosed in humid atmospheres, even when the metal is first given a pretreatment with two pack polyvinylbutyral-chromate-phosphoric etch primer, which usually ensures a high standard of performance on outdoor exposure. The volatile acids, generated by the oxidative curing of the drying-oils present, instead of escaping into the free atmosphere as it does on exposure, builds up to corrosive levels and attacks the metal surface. Fig. 6.7 shows a typical example of a painted zinc-coated steel surface stored in a warm humid atmosphere where the paint has flaked away under a flap, and the zinc of the underlying galvanizing has corroded and attack of the base steel has begun. Paint on steel is less likely to suffer in this way from the acids generated within the paint.

Paint gives good protection to steel against moderate levels of organic acid vapours but protection eventually breaks down in higher concentrations. Fig. 6.8 for instance shows extensive rusting of steel fittings and painted steel surfaces by high concentrations of acetic acid evolved from the wooden package.

The steel rocket motor in Fig. 6.8 had been phosphated and painted with

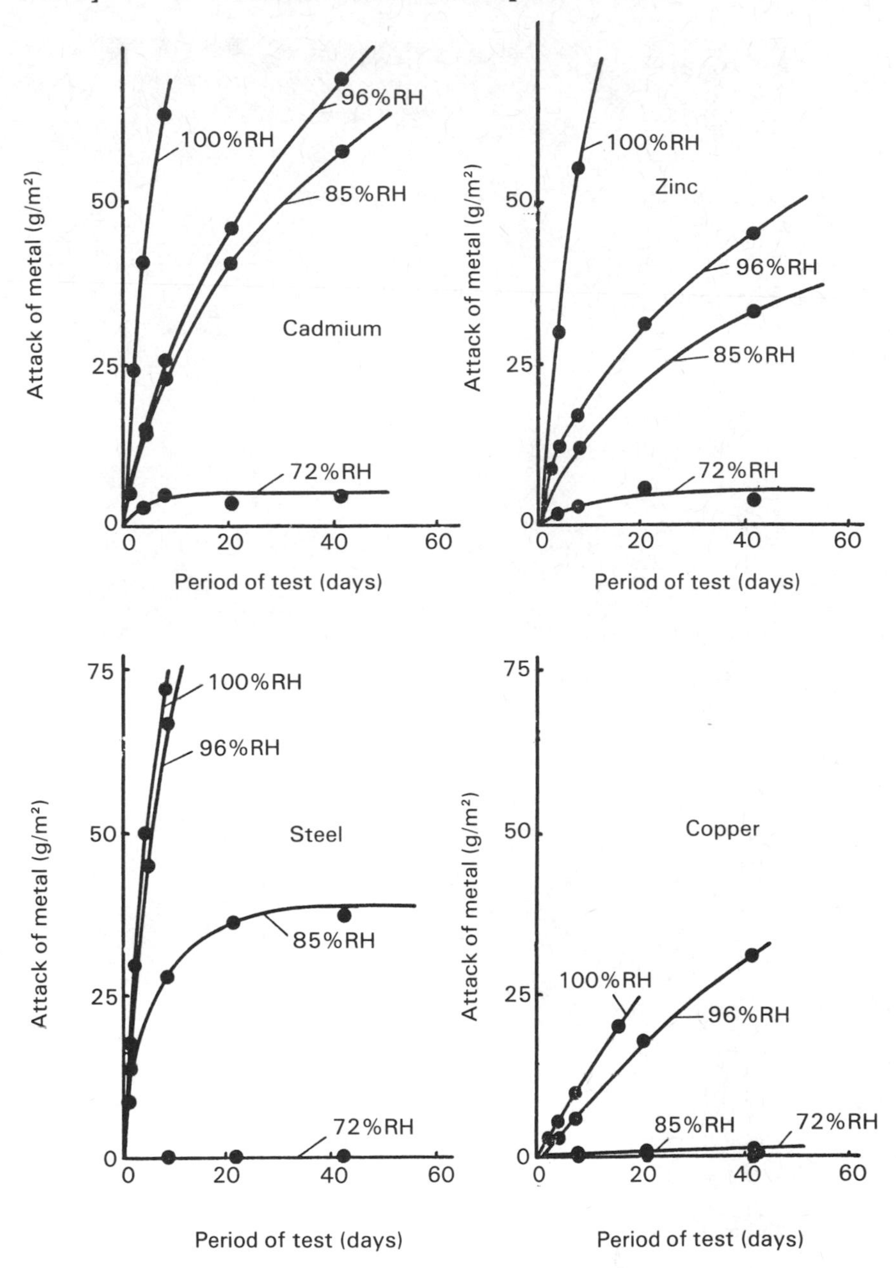

Fig. 6.6 — Effect of relative humidity on corrosion in 50 ppm acetic acid vapour. (© Controller, Her Majesty's Stationery Office 1986.)

three coats of stoving paint and had then been subjected to an environmental test simulating tropical exposure (ISAT A), in its wooden package. Not only were the bare steel fittings attacked so vigorously that they were scarcely recognisable but the painted steel surfaces were also badly rusted.

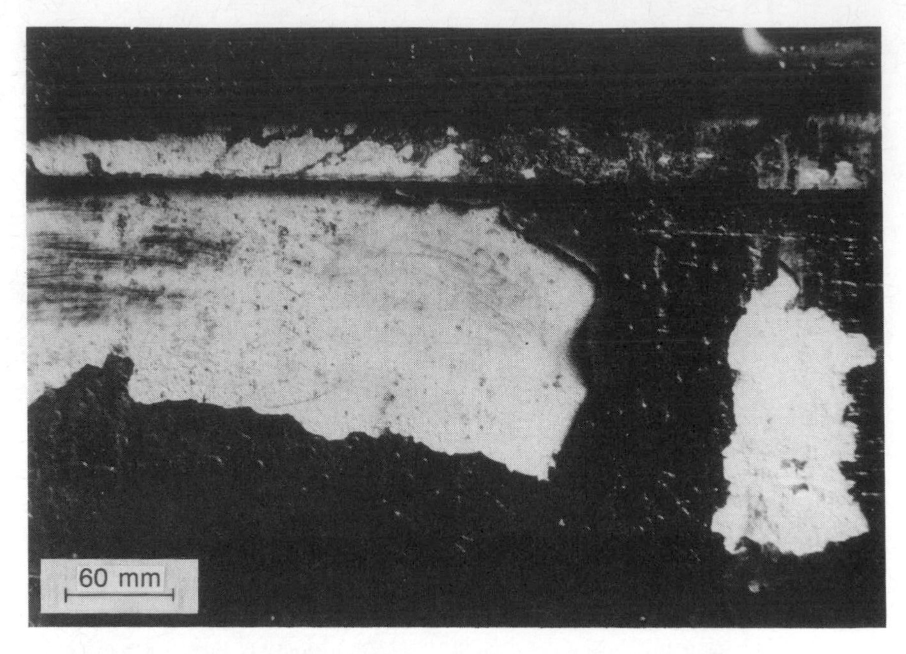

Fig. 6.7 — Flaking of oleo-resinous paint from zinc confined in a damp atmosphere. (© Controller, Her Majesty's Stationery Office 1986.)

Fig. 6.8 — Painted steel rocket motor in wooden box corroded on environmental test (ISAT A). (© Controller, Her Majesty's Stationery Office 1986.)

Paint on tin coatings on steel gives a high level of protection both against organic acid vapours in humid enclosures and on outdoor exposure.

From among the temporary protectives, solvent-deposited lanolin films give little protection against organic acid vapours, as the film is thin and the volatile organic acids appear to dissolve readily in the long-chain organic acids of the coating. Solvent-deposited bitumen-based coatings, or films of petrolatum, give better protection. Protection improves as coating thickness increases.

Chromate passivation on both zinc and cadmium provides good initial protection against vapour corrosion attack, as it does against outdoor environments, but the films are so thin and easily damaged that the extra degree of protection they afford cannot be relied on for prolonged exposure.

HYDROGEN EMBRITTLEMENT OF STEEL BY VAPOUR CORROSION

High strength steels are susceptible to failure at sustained stress levels well below their tensile strength if they contain hydrogen (see Chapter 5). Any corrosion that may give rise to hydrogen is therefore particularly undesirable with these steels. The favoured cathodic reaction during atmospheric exposure is oxygen reduction, but in crevices, under paint films and in confined air spaces oxygen may become exhausted, and hydrogen may be released. The effect is illustrated in Fig. 6.9 which shows the pressure changes due to corrosion of a high strength steel in a confined atmosphere with (a) acetic acid vapour, (b) formic acid vapour and (c) the surface contaminated with synthetic sea water.

It can be seen from Fig. 6.9 that after the pressures fall corresponding to exhaustion of oxygen (drop of 0.2 atmospheres) there is a steady increase of pressure indicating hydrogen release when formic or acetic acid is present; with the neutral conditions of salt corrosion there is no such pressure rise, and therefore little apparent risk of hydrogen embrittlement.

WOOD AS A SOURCE OF CORROSIVE VAPOURS

Some timbers, in the as-felled condition, contain free acetic acid; oak and sweet chestnut are two noted examples and both have a long history as sources of corrosive vapours. Most other commercially exploited timbers contain much less free acetic acid than oak and sweet chestnut but all woods have acetyl groups incorporated into the hemicellulose. Arni, Cochrane & Gray (1965), in a study of the release of acid from wood under warm moist conditions, examined samples of heartwood and sapwood from trees felled in the autumn, winter and spring. They showed that oak and sweet chestnut contained about 0.2% free acetic acid when felled; most of the other woods examined contained no free acetic acid. Spring-felled woods generally released acetic acid more rapidly than the corresponding winter- or autumn-felled specimens; there was little difference in release of volatile acid from heartwood or sapwood. Acetyl groups in the wood were hydrolysed at

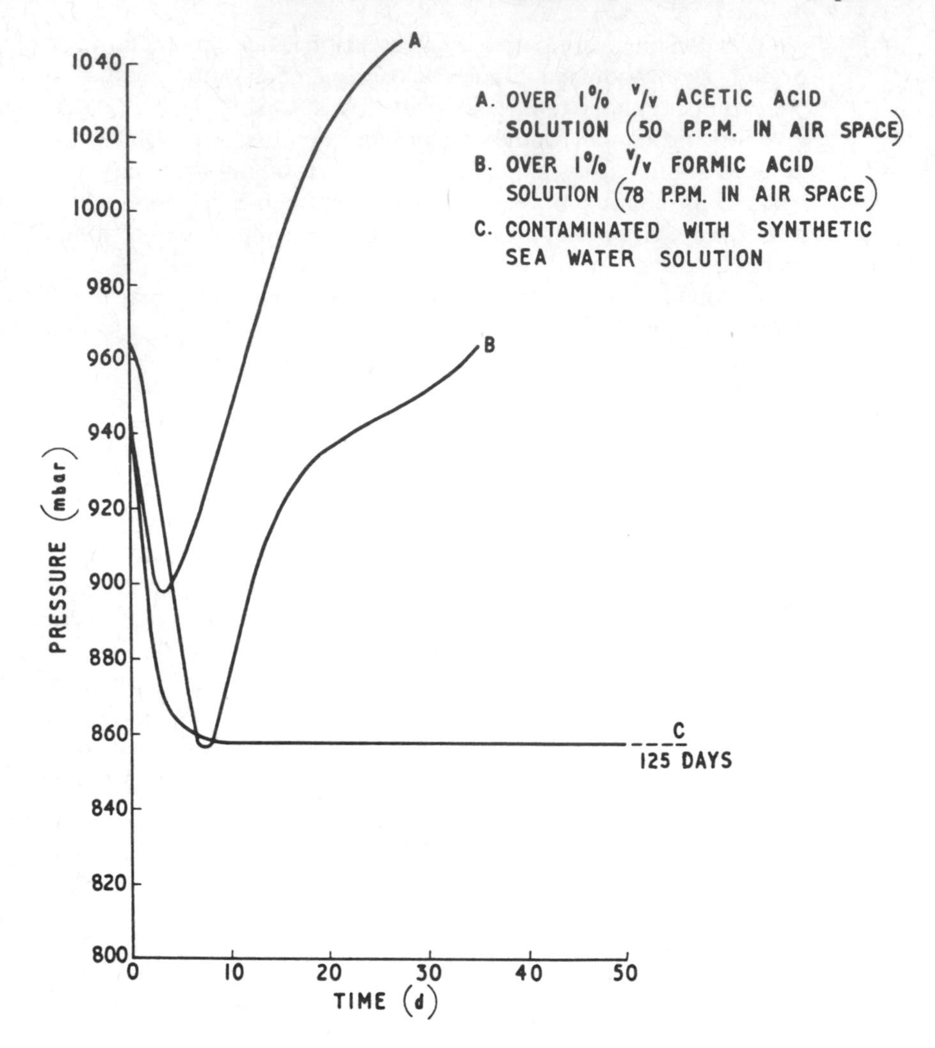

Fig. 6.9 — Pressure variations due to corrosion of high strength steel in a closed container. (© Controller, Her Majesty's Stationery Office 1986.)

appreciable rates at 30°C but more rapidly at 48°C. Table 6.5 gives the acetyl content of a series of freshly felled woods, that remaining after 39 weeks of storage at 48°C and the quantity of acetic acid released (from Arni, Cochrane & Gray (1965)).

Although oak has a lower content of combined acetyl than most other woods, hydrolysis to release acetic acid occurs more readily. This and the free acetic acid already present account for the high corrosivity of oak. Arni, Cochrane & Gray (1965) also observed that acid release was autocatalytic in its early stages, and that woods initially free of acetic acid, such as Norway spruce, experienced a latent period before hydrolysis started.

There have been many instances of corrosion by woods, other than oak

Table 6.5 — Acetyl and volatile acids in heart woods as-felled and after ageing at 48°C (reproduced with permission from Arni, Cochrane & Gray (1965))

Wood	As-felled		After 39 at 48°C and 100% RH	Volatile acid evolved[b] at 48°C for 26 weeks	
	Volatile acid (%)[a]	Acetyl (%)[a]	Acetyl (%)[a]	Formic acid	Acetic acid
Sweet chestnut	0.04	3.6	0.6	102	759
Oak	0.27	1.9	0.4	79	630
Wych elm		5.0	4.5	29	402
Douglas fir		1.1	0.5	62	565
Norway spruce		1.3	0.7	0.5	32
Beech				4	37

a g per 100 g of oven-dried wood
b Parts per million (w/w) on weight of wood as-felled

and sweet chestnut, reported over the past fifty years in temperate climates, and in practically every instance the wood has been subject to a prior heat treatment, such as kiln drying from the green state or after aqueous rot-proofing or fire-proofing treatment, or in steam bending or resin bonding.

A typical early instance, reported by Burns & Freed (1928), involved corrosion of lead-covered cables carried in a wooden duct made from Douglas fir. The corrosion was caused by acetic acid from the wood, which had been subjected to vacuum impregnation with creosote at temperatures of 71°C–93°C over a period of 20 hours. The corrosivity of the ducting was overcome in this instance by treatment with ammonia vapour so that the acetic acid was converted to the non-volatile ammonium salt.

Schikorr (1961) reported corrosion of steel and brass clock mechanisms caused by acetic acid released from the wood of their plywood cases. He found that in some instances steel components had rusted before the clocks had left the manufacturer's warehouse. Some protection was obtained from wrapping the clockwork movement with paper impregnated with the volatile inhibitor dicyclohexylamine nitrite (DCHN). A lacquer heavily loaded with chalk (calcium carbonate) also reduced the corrosion.

Clarke & Longhurst (1961) showed clearly how high temperature treatments render wood corrosive. They used two woods in their experiments: (a) spruce planking, and (b) resin-bonded 4 mm plywood. Pieces of each of these woods were rot-proofed by aqueous treatments. All the treated woods caused severe corrosion of zinc when tested in a test bottle similar to that shown in Fig. 6.2; the loss of metal was equivalent to corrosion of from 12 μm to 65 μm thickness in three weeks. Untreated timber and plywood caused no corrosion in the same test. These workers, suspecting that the forced drying after impregnation was responsible for the corrosion observed, wetted the woods with distilled water and then dried specimens in the laboratory under conditions similar to those use in the kiln drying, viz.

Stage	Period (days)	Temp. (°C)	RH in kiln (%)
1	7	65	80
2	7	70	64
3	7	75	34

The specimens when enclosed with zinc specimens gave the same rapid corrosion as did the rot-proofed woods.

Wood of any type represents a risk of corrosion to metals enclosed with it — a risk which is difficult to wholly avoid. The best answer lies in the removal of the wood; however in this connection the ubiquity of wood in artefacts, as packing pieces, space fillers, handles and resin-bonded composites, often makes is difficult to ensure its complete absence. Fig. 6.10 for instance shows a switch mechanism which corroded shortly after its assembly within an enclosure formed by an aluminium casting. The zinc-

Fig. 6.10 — Enclosed swtich corroded after six months of storage by vapours from wood-flour filled phenolic moulding. (© Controller, Her Majesty's Stationery Office 1986.)

plated steel and brass items of the switch were vigorously attacked. The corrosive impurity was shown to be mainly acetic acid derived from the black phenolic moulding, visible in the figure, which contained a wood-flour filler. The filler would have been heated to about 170°C during manufacture of the phenolic moulding. A ceramic filler was substituted and corrosion no longer occurred.

Interesting and regrettable examples of the unwise use of wood for storage of metals are constantly visible in the continuing use of wooden display and storage cabinets in many museums where irreplaceable metal objects are stored for what is expected to be an enduring life. The instances of disintegration of sea shells stored in this way (the calcium carbonate of shells reacts more readily than metals, although only in stoichiometric proportions) and of 'bronze disease' are sometimes seen as unfortunate and exceptional instances, but in terms of storage over generations they can more properly be regarded as high probability events for all metals stored in wooden cases; synthetic adhesives often add to the risk. (BS 5454 mentions this problem in the vapour corrosion of the lead seals on old documents.)

CORROSIVE VAPOURS FROM PLASTICS AND PAINTS

An important advantage of synthetic polymers over natural materials is their general inertness and resistance to deterioration under ambient conditions. This inertness of the bulk of the polymer may be deceptive if it leads to the assumption that because the plastic remains apparently unchanged it will have no effect on surrounding materials. Practically all plastics contain residues from their manufacture which are volatile and are subsequently released into the surrounding atmosphere. Others degrade slowly or depolymerise. Ageing reactions may often be induced or accelerated by high temperatures, ultra-violet radiation, oxidation, microbiological action, and contamination with acids, alkali or oxidising agents and sometimes metallic salts. Commercial polymers are generally built on a carbon skeleton and their breakdown usually involves oxidation by reaction with the atmosphere, so that most volatile fragments are rich in carbon and oxygen. The most frequently encountered breakdown products are carbon dioxide (CO_2), carbon monoxide (CO), formaldehyde (HCHO), formic acid (HCO_2H), acetaldehyde (CH_3CHO) and acetic acid (CH_3CO_2H). In some polymers the presence of chemical groups on side chains may give other volatile acids: thus polyvinyl chloride (PVC) may give rise to hydrogen chloride (HCl), and nitrocellulose oxidises to give a mixture of oxides of nitrogen which with water give nitrous (HNO_2) and nitric (HNO_3) acids. Cross-linked systems may give side reactions during curing which result in more complex fragments; thus drying-oils during cure give not only considerable quantities of formic acid and formaldehyde but also higher fatty acids and aldehydes including propionaldehyde, hexanal, valeral and heptanal. Traces of solvents may also be retained, especially in lacquers and varnishes, and these can be a corrosion hazard. Fillers in plastics are a possible source of corrosive vapours.

Three phases may be distinguished in the life of plastics, in each of which corrosive vapours may be evolved: manufacture, the mature life of the plastic, and final degradation leading to loss of physical properties. Problems during manufacture although important are outside the scope of this book. The second category is of major concern and most of this section is devoted to discussing problems which arise from vapours released by new plastics. Corrosive vapours may also arise from ageing or heating plastics and this subject is discussed later in this chapter. Some of the particular effects associated with biological activity are summarised in Chapter 9. Usually the quantity of vapours evolved is insufficient to give rise to troublesome concentrations in well-ventilated atmospheres but in the confined atmospheres within a package or store-room concentrations often become sufficient to cause rapid corrosion.

Having alerted the reader to the possible hazards from polymeric materials, balance may perhaps be given by an assurance that many polymeric materials are not liable to evolve corrosive vapours and are safe to use in enclosures with metals. However even this assurance must be tempered with a note of caution. Each plastic named is in fact a generic title of a family of materials and even within one detailed formulation individual manufacturers introduce variations in stabilizers, end-stoppers, fillers, and other additives, and use ingredients of varying chain length and with different levels and types of impurities. Formulation and processing vary in detail from one manufacturer to another, and even from batch to batch from the same manufacturer; sometimes variations are introduced to give improvements in the product or to economise on the cost of manufacture so that formulations are changed, often without warning to the customer. These variations may not change, or may even improve, the physical properties of the material which are important in its primary function, although they may be very significant in giving rise to by-products. But the results quoted in this chapter are based on sufficient experience with numerous samples of each formulation to be reasonably confident of the classification given. Where variations are known to arise the range of possibilities is indicated. More detailed assurances can only be derived from tests of the type illustrated in Fig. 6.2 and discussed later.

Formulations which have not been known, in bulk (NB excluding glues, jointing compositions and paints), to give rise to corrosive vapours in practice and have given no corrosion on zinc or steel specimens in tests at 100% RH in either warm (30°C) or hot (60°C) wet atmospheres, are:

- Polythene, polypropylene;
- Polymethyl methacrylate, linear polyesters, diallyl phthalate (DAP), polystyrene, acrylonitrile-butadiene-styrene (ABS), polycarbonate, polyphenylene oxide, polyhydroxy ether, polysulphone;
- melamine formaldehyde, polyurethanes (when correctly formulated, fully cured and free of fillers);
- casein, ethyl cellulose.

Table 6.6 lists the plastics and rubbers which are known to have been a cause of corrosion. This list is not exhaustive as it is limited to materials for which test results are available. It is also based on a comparatively coarse test; most polymers for instance, even those listed above as inert, give sufficient volatiles to adversely affect integrated circuits (see for instance the work of Licari & Browning (1967) on studies of possible failure modes of Minuteman 2, which describes the degradation of the life and reliability of transistors and integrated circuits exposed to vapours from plastics), but an account of these effects is beyond the scope of this book.

A typical failure caused by a cold-curing rubber is shown in Fig. 6.11. In

Table 6.6 — Plastics and rubbers giving corrosive vapours (reproduced with permission from Donovan & Stringer (1971))

Material	Severity of corrosion	Volatiles/remarks
Thermoplastics		
PVC and other chlorinated plastics	Non-corrosive at ambient temperature (but see column 3); moderately to very corrosive at 70°C	HCl. May become corrosive at lower temperature if irradiated with u.v. or in contact with some chemicals
Fluorinated plastics e.g. PTFE	Non-corrosive up to moderate temperatures; very corrosive above about 350°C	HF and F_2
Nitrocellulose	Slightly to very corrosive	Nitrogen oxides evolved progessively on ageing
Nylon 6	Nylon 6 corrosive; other nylons non-corrosive	Acetic acid; from acetate end-stoppers
PVA (polyvinyl acetate)	Non-corrosive to very corrosive	Acetic acid; quantity depends on stabilisers and inhibitors, and extent of prior hydrolysis
Cellulose acetate	Slightly to moderately corrosive	Acetic acid; release increases with age, and acid conditions
Polyacetals		
(a) Homopolymer	Slightly corrosive at ambient temperature; more corrosive above 40°C	Acetic and formic acids; acetic acid used as end-stopper
(b) Copolymer (formaldehyde and 10% ethylene oxide)	Corrosive above 45°C	Formic acid
Thermosetting resins		
Polyester resins		
(a) Cold-cured resins	MEKP catalysed systems are very corrosive; others slightly to moderately corrosive	Acetic and formic acids. Catalyst is main cause but diethylene gylcol gives more corrosive resins than propane-1,2-diol
(b) Hot-cured polyester resins	Non-corrosive to moderately corrosive	Formic acid
Phenolformaldehyde		
(a) Two-stage resins (via novolaks)	Moderately to very corrosive	Formaldehyde and ammonia
(b) One-stage cure (via resols)	Non-corrosive (if fully cured and with inert fillers)	
Epoxides		
Cold-cured epoxides	Non-corrosive to slightly corrosive	Amines from amine-catalysed resins and cresol from some low exotherm resins
Rubbers, elastomers and adhesives		
Natural rubber		
(a) Non-vulcanised	Slightly corrosive on prolonged exposure	Formic and acetic acids from oxidation and depolymerisation
(b) Vulcanised	Slightly to moderately corrosive	Hydrogen sulphide and sulphur dioxide
Synthetic rubber		
(a) Chlorinated	Non-corrosive but corrosive vapours evolved above about 100°C	HCl. Hypalon may give sulphur dioxide
(b) Polysulphide rubber (cold curing)	Moderately to very corrosive	Formic acid; the catalysts are peroxides
Silicone polymers	Non-corrosive to very corrosive	Acetic and formic acids; some single-pack sealants cure by acetoxyhydrolysis and are very corrosive; some two-pack systems evolve formic acid and are very corrosive; others are inert
Phenolic and urea–formaldehyde glues	Slightly to very corrosive	Formaldehyde. phenol, ammonia and HCl. Acid catalysts are often used in cold setting formulations. Volatiles released in curing are absorbed in bonded materials

the 'sealed' optical assembly shown in Fig. 6.11 cadmium plating was seriously corroded and phosphated steel had rusted before the instruments had reached service. The corrosion was caused by formic acid released by a cold-curing rubber used to mount the lenses. The rubber had been cured by addition of a lead peroxide catalyst which presumably caused some oxidative breakdown of the rubber.

Fig. 6.11 — Internal components of optical mechanism corroded by vapours from cold-curing rubber. (© Controller, Her Majesty's Stationery Office 1986.)

Little work has been reported on formulating polymers to reduce the evolution of corrosive vapours, with the notable exception of cross-linked polyesters, which have been the subject of particularly intensive investigations since they have been widely used with glass fibre reinforcement in fabricating large packages, radomes, boxes for switchgear and many large structures which often contain metallic components. An electronic assembly which suffered extensive corrosion, from vapours derived from a combination of a polyester package and PVA-bonded hair used as cushioning, is shown in Fig. 6.12. The electronic assembly corroded, although it was sealed within a thick polythene inner wrap.

Work on the causes of corrosion by polyesters carried out at Yarsley Laboratories by Cawthorne, Flavell & Ross (1966), and by Cawthorne, Flavell & Pinchin (1969), was based on experiments using a standard formulation:

Phthalic anhydride	1.0 g mole
Maleic anhydride	1.0 g mole
Diethylene glycol	1.2 g mole

Propane-1,2-diol 1.2 g mole
Compounded with 30% (w/w) styrene.

Fig. 6.12 — Electronic assembly corroded by vapours from polyester package and PVA-bonded hair. (© Controller, Her Majesty's Stationery Office 1986.)

Using this standard formulation it was shown that the corrosive vapours resulted both from decomposition of the peroxide catalyst and from oxidation of styrene by the peroxide; resins cured without peroxides, by gamma radiation for instance, were not corrosive. The corrosivity of polyesters varied with the type and amount of catalyst and accelerator used. Using 2% additions of a range of peroxide catalysts gave resins which varied in the extent of corrosion caused by an order of magnitude. The peroxides used in order of corrosivity, with the attack on zinc (g/m^2) caused by resins cured with the peroxides, in a 21-day test at 100% RH, was: *t*-butyl hydroperoxide (56), 2.4 dichloro benzoyl peroxide (41) *t*-butyl perbenzoate (20), methyl ethyl ketone hydroperoxide (17), dicumyl peroxide (14), cumene hydroperoxide (12), cyclohexanone peroxide (8), lauroyl peroxide (8), benzoyl peroxide (7).

The corrosivity of the resins also varied with the cross-linking glycol, with propane-1,2-diol giving substantially less corrosion than diethylene glycol. Addition of sodium bicarbonate was shown to reduce the quantity of acid vapours released. Full details of this work are best read from the original papers but Table 6.7. summarizes results of corrosion tests carried out on polyester resins formulated with propane-1,2-diol and catalysts chosen from those giving least corrosion.

The resin formulations tested showed a decrease in corrosivity compared with unmodified formulations; in other experiments the standard resin with 2% methyl ethyl ketone peroxide and 2% cobalt naphthenate gave 230 g/m^2 attack of the zinc in the test at 30°C and 100% RH for 35 days. The addition of sodium bicarbonate to the least corrosive formulations reduced the corrosion of zinc to low levels although it did not prove possible to produce a fully non-corrosive cold-curing formulation suitable for commercial use. The formulation with lauroyl peroxide as catalyst was non-corrosive but required a post-curing treatment at 50°C.

Polyvinyl acetate is often used as a bonding agent in the glass mats used as reinforcement for polyesters; it is also frequently used as a release agent on the surfaces of the moulds used with these composites. PVA from either source gives corrosive resins and is therefore to be avoided if the composite is to be used near metals.

Andrew, Stringer & Weighill (1977) evaluated the formulations developed by Cawthorne and his co-workers (1966, 1969). They fabricated glass-reinforced polyester packages from the preferred resins and showed that they were essentially non-corrosive to a wide range of metals, even in hot tropical conditions.

PAINTS

Paints have often been reported as a source of corrosive vapours. Under-cured phenolformaldehyde resin paint and air-drying oleoresinous paints are both well established as corrosion risks. A variety of organic acids and aldehydes are released during the early stages of cure of air-drying oleoresinous paints. After 10 to 14 days curing the main volatile evolved is formic acid which give continues to be released over a prolonged period at a rate sufficient to a corrosive concentration in an enclosed atmosphere. Corrosive vapours are also evolved from stoving oleoresinous paints for some time after cure, although little work has been reported on these materials.

Table 6.8, from work by Donovan & Moynehan (1965), gives the results of corrosion tests on a range of air-drying paints and varnishes applied to glass slides and exposed with zinc coupons. It can be seen that the drying-oil based paints gave high levels of corrosion; this is in accord with the many failures attributed to these paints. Heavy corrosion was also caused by the acrylic paint. The corrosion by the acrylic paint was shown to be due to acetic acid derived from cellosolve acetate used as a solvent, which explains why the thicker film caused so much more corrosion since it would have retained proportionately more solvent. Corrosion by the nitrocellulose lacquer was later shown to be due to the same effect although it proved more difficult to identify the source of contamination since the solvent was the apparently innocuous cellosolve, but it was shown to contain 1% cellosolve acetate, which was within the impurity limits allowed by the specification of the solvent.

In the tests shown in Table 6.8 the area of paint was the same as that of the exposed zinc (both had surface areas of 25 cm^2). In practice the area of

Table 6.7 — Corrosion tests on cold-cured 'non-corrosive' polyester resins (reproduced with permission from Donovan & Stringer (1971))

Catalysts (% w/w)	Sodium Bicarbonate (% w/w in resin)[a]	Weight loss of zinc (g/m^2)	
		35 days, 30°C 100% RH	56 days (ISAT A) (cycling to 60°C)
Cyclohexanone peroxide (2%) and cobalt naphthenate solution (2%)	NIl[b] → Nil[b]	44	107
	0.05[b] → 0.5[b]	13	15
Benzoyl peroxide and *NN*-dimethylaniline solution (2%)	Nil	47	190
	0.5[b]	3	99
	0.5[c]	6	62
Lauroyl peroxide (2%) and *NN*-dimethylani line solution (4%)	Nil	9	11
	0.5[b]	3	4
Control — no resin		4	8

a Resin based on propylene-1,2-diol cured at room temperature except when lauroyl peroxide was used when curing was at 50°C for 4 hours
b 325 mesh
c Spray dried

Table 6.8 — Corrosion of zinc by vapours from paints (paints air-dried for 24 hours; 3 weeks, 30°C, 100% RH)

Medium	Film weight (g/m^2)	Corrosion of zinc surface (μm)
Linseed oil alkyd	50	7.8
Dehydrated castor oil modified	50[a]	2.0
epoxy-ester	120[a] → 120	5.2
Vinyl toluenated modified	20	1.6
epoxy-ester	70[b]	5.0
Chlorinated rubber	50	0.4
Nitrocellulose	40	2.7
Polyurethane (2 pack)	110	0.1
Polyamide-cured epoxide	60	0.1
Amine adduct-cured epoxide	50	0.1
Solvent acrylic paint	50	5.0
	100	27
Solvent acrylic lacquer	40	0.1
Control	None	0.2

a thinned with xylol. *b* two coats.

paint is usually very much greater than that of the metal at risk so that relatively more acid vapours are available for attack.

The compositions of the volatiles evolved from the oil-drying paints were investigated by passing air at a rate of two litres a minute through a Winchester bottle, the inside of which had been coated with the paint. It was shown that for the first few days a mixture of all the lower saturated carboxylic acids up to caproic acid was evolved but the major proportion was formic acid. A mixture of aldehydes was also identified, these corresponded closely to the predictable products from the oxidative breakdown of the unsaturated fatty acid residues. Thus linseed oil-based materials gave mostly propionaldehyde (propanal) together with lesser amounts of hexanal and an unsaturated aldehyde (thought to be hex-2-enal). Dehydrated castor oil-based materials gave a mixture of hexanal, valeral and heptanal; tung oil gave mostly valeral. After a few days the proportion of higher acids and aldehydes reduced rapidly and the major volatile evolved over longer periods was formic acid.

Fig. 6.13 shows the rate of evolution of formic acid from air-drying paints over long time-periods. All the drying-oil based paints tested continued to evolve formic acid up to 100 days after application, and the three tested for longer periods continued to evolve formic acid for up to a further 100 days with little change in the rate of evolution.

One paint, after the 200-day test of Fig. 6.13, was stored for a further 200 days without air flow, and then the air flow was restarted when the further evolution of formic acid occurred, as shown in Fig. 6.14.

Force drying at 70°C for 45 minutes reduced the evolution of formic acid over the first 60 days but had little effect on its longer term evolution. Results of corrosion tests on oleoresinous paints which had been allowed to cure for up to six months are shown in Fig. 6.15.

Calcium plumbate-containing paints, formulated as primers for direct application to zinc, are the only air-drying oleoresinous paints found by the writer to be non-corrosive (at 30°C and 100% RH); this is presumably because the alkaline calcium plumbate reacts with any acids released. This is almost certainly the same mechanism which, at least in part, makes them suitable for use on zinc, since it is the reaction of the zinc metal with the free organic acids from the curing of the drying oils which induces failure with other oloeresinous paints.

Polyvinyl acetate-based emulsion paints are potential sources of acetic acid vapours but in tests at 100% RH and 30°C undertaken by the writer corrosion varied from very heavy to none. The variation is thought to depend on both the degree of prior hydrolysis of the acetate groups, the efficiency with which traces of free acetic acid are removed in manufacture, and the variation in the type and concentration of stabilizing and inhibiting agents in the paints.

In summary the position of paints as a source of corrosive vapour is:

— Drying-oil and semi-drying-oil based paints, and air-cured paints con-

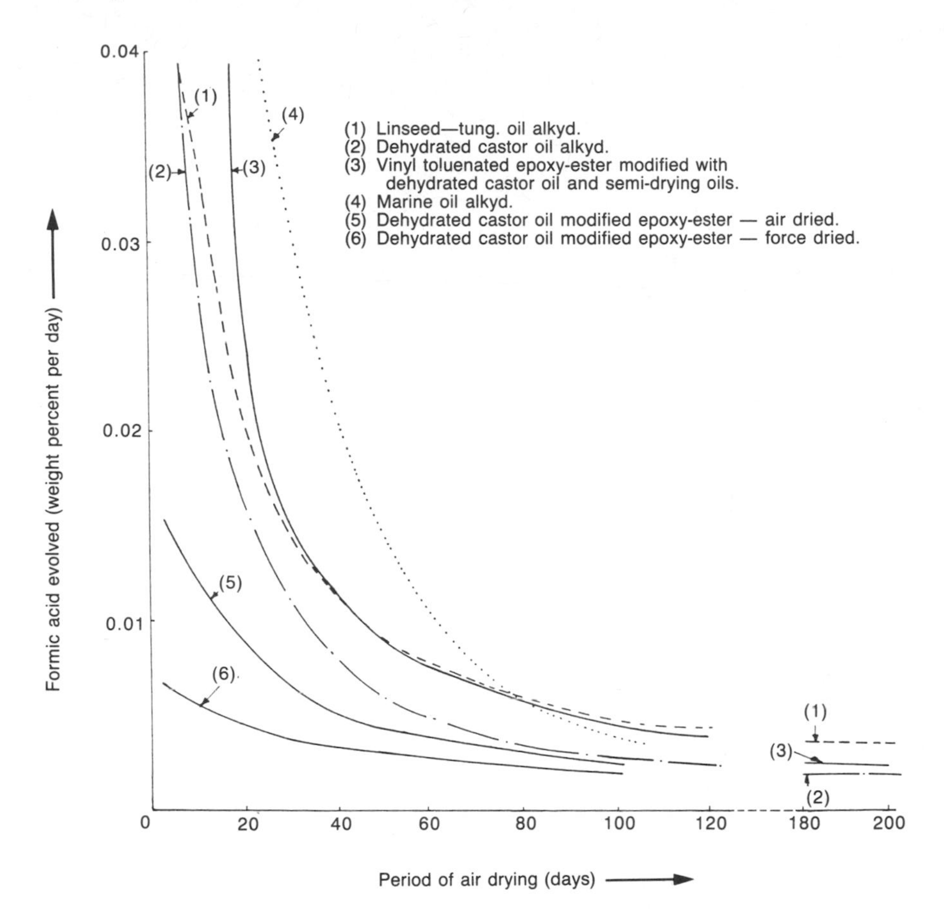

Fig. 6.13 — Formic acid release from oleoresinous paints at 20°C. (© Controller, Her Majesty's Stationery Office 1986.)

taining esters of volatile acids in the solvent are likely to give rise to serious vapour corrosion.

— Epoxides, polyurethane (two-pack), solvent acrylic, chlorinated rubber and nitrocellulose-based paints do not appear to cause vapour corrosion provided that the paint solvents do not contain esters of volatile acids and have not been modified by drying-oils or semi-drying-oils. But both chlorinated rubber and nitrocellulose paints must be considered with reservation for long-term exposure since both may degrade, evolving respectively hydrogen chloride and oxyacids of nitrogen, which are corrosive.

— Polyvinyl acetate emulsion paints are sometimes very corrosive; those which are not initially corrosive are almost certainly a risk in the longer term. Polyvinyl acetate is also widely used as a binder and adhesive, where it has frequently been found to be a prolific source of acetic acid vapour.

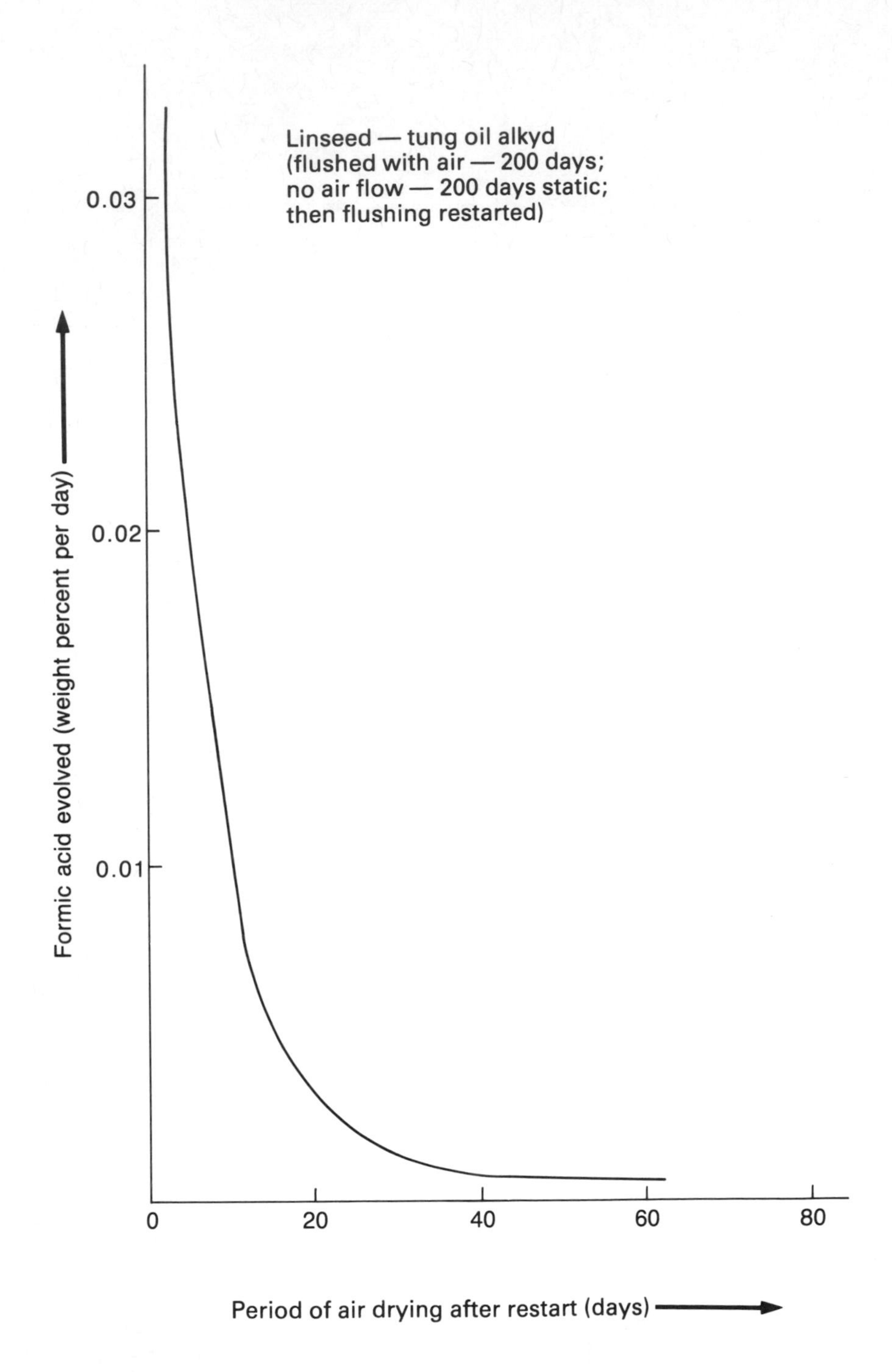

Fig. 6.14 — Rate of formic acid release from an air-drying paint after 400 days curing. (© Controller, Her Majesty's Stationery Office 1986.)

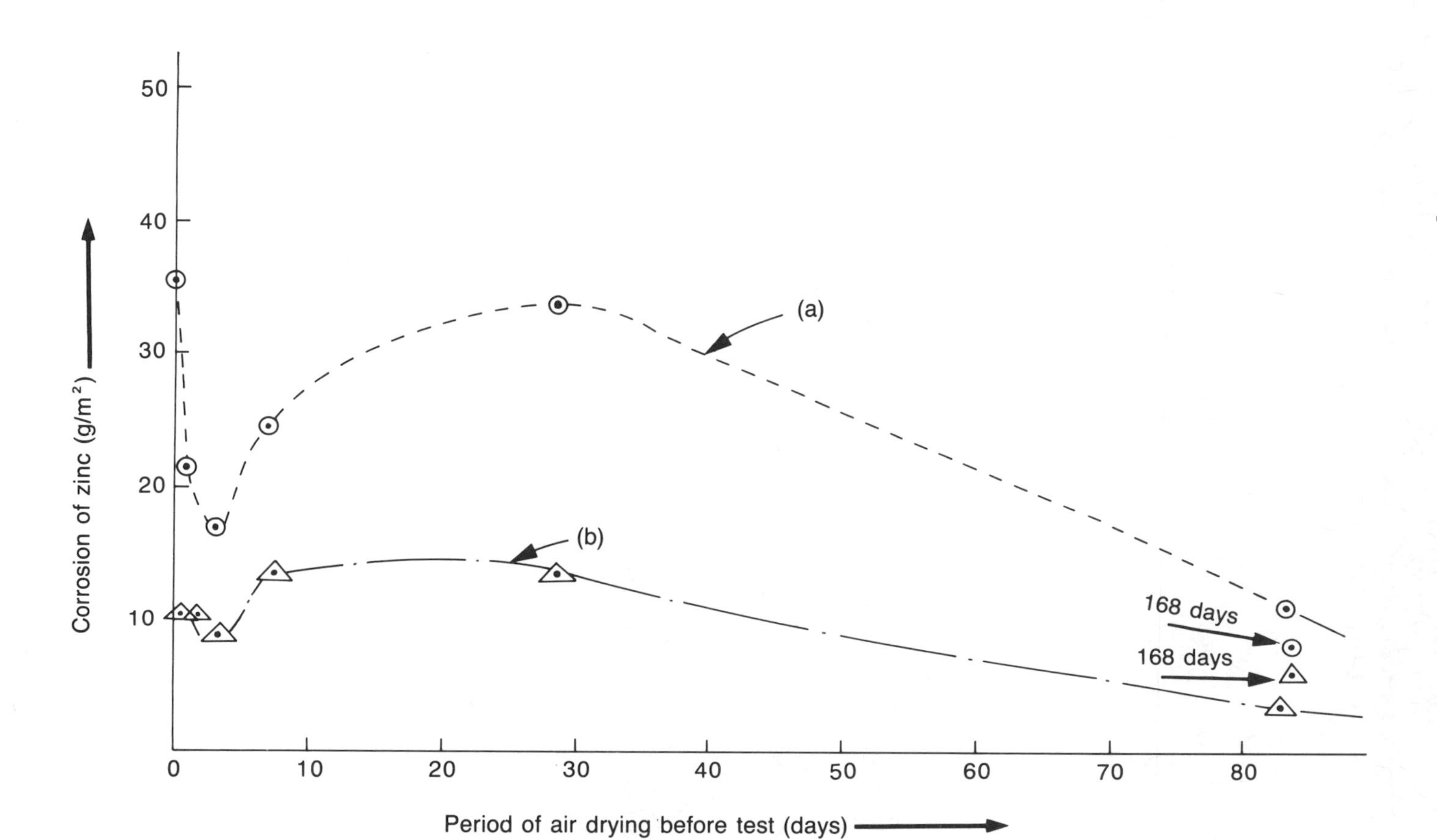

Fig. 6.15 — Corrosion of zinc by vapours from air-drying paints after extended curing. (a) Linseed oil alkyd. (b) Modified epoxy-ester (vinyl toluenated epoxy-ester modified with dehydrated castor oil and semi-drying oils). (© Controller, Her Majesty's Stationery Office 1986.)

— Formaldehyde resin-based paints and lacquers may be very corrosive if they are not fully cured.

OTHER SOURCES OF CORROSIVE VAPOURS

Adhesives have often been cited as sources of corrosion (see for instance Rance & Cole (1958) and Knotkova-Cermakova & Vlckova (1971)) and *ad hoc* testing has established that many of the synthetic resin adhesives evolve volatile acids during curing. The vapours evolved are often absorbed by the materials being bonded, to be slowly released during later service. Many plywoods made with synthetic glues (urea–formaldehyde, phenol–formaldehyde, melamine–formaldehyde and resorcinol–formaldehyde) have been shown to be corrosive. If the plywoods are hot cured then acetic acid from the wood forms part of the corrosive volatiles and the extent of acid release from the veneer depends on the curing temperatures and type of wood but the glues also make a substantial contribution.

Heating during the manufacture may be avoided by using cold-curing adhesives but some cold-cured plywoods have been among the most corrosive systems encountered by the writer — one in particular was a two-component rapid-curing adhesive in which one of the components was formic acid. In experiments carried out by the writer, a standard veneer, which on its own gave little corrosion in tests with zinc at 30°C , was fabricated into plywoods using a range of cold-curing resins in a controlled manufacturing environment. A urea formaldehyde/ melamine formulation gave heavy corrosion; others gave lesser degrees of corrosion, with a phenol formaldehyde-bonded veneer giving very little corrosion. When the same bonded veneers were tested under cycling conditions which included exposure at temperatures up to 60°C, all were corrosive, again showing that although the hazard of vapour corrosion from wood can be kept low in temperate climates by careful selection of the wood, and by controlling processing conditions, in hot tropical conditions any wood products are probable sources of corrosive vapours.

Wood and drying-oils are two sources of corrosive vapours which are widely used in the formulation of composite boards, synthetic composites, moulding powders and fillers, or as packaging materials or in the construction of storage areas. Chipboard, hardboard and wood-flour filled potting resins are all typical compositions which release acetic acid. Linoleum, linseed oil putties and other jointing compositions frequently contain drying oils and evolve volatile organic acids throughout their life through the same mechanism as oleoresinous paints but at an even greater rate. An early reported observation of vapour corrosion was from within a plant manufacturing linoleum (Angew, 1915).

An example of corrosion by a drying-oil is shown in Fig. 6.16. The corroded steel components in Fig. 6.16 were from a composite assembly, the wooden parts of which had been treated with linseed oil. The steel was protected with a thick oil (PX-4) and the assembly was wrapped in waxed

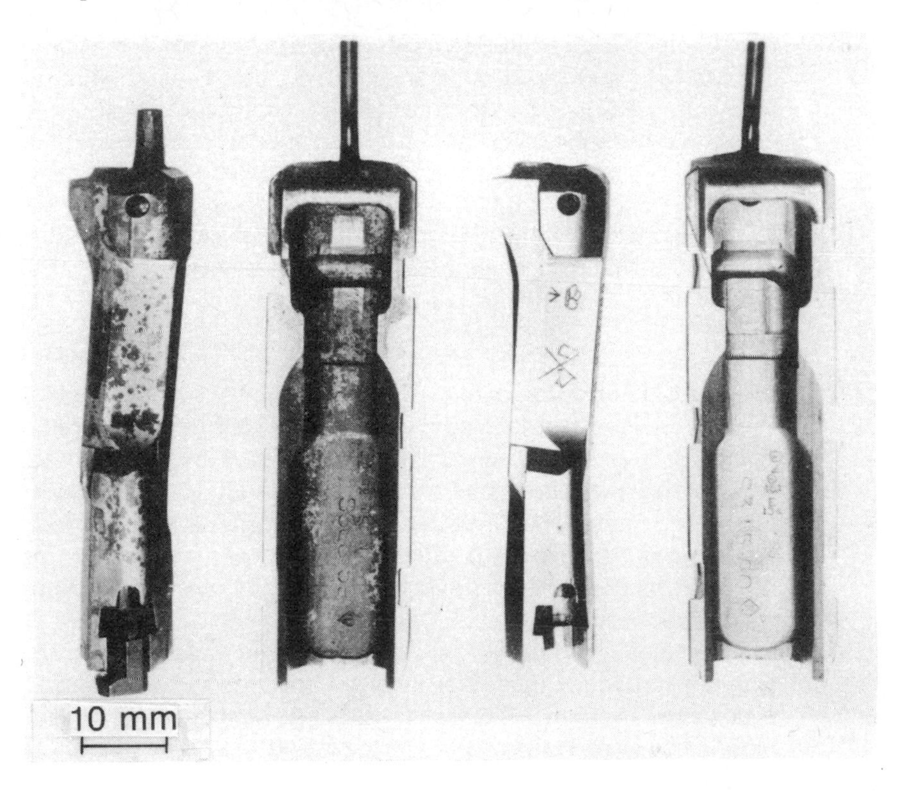

Fig. 6.16 — Corrosion of steel by vapours from linseed oil. (© Controller, Her Majesty's Stationery Office 1986.)

paper but corrosion occurred within a year when the components were stored in a humid tropical atmosphere.

Other sources of organic acid vapours include oil of turpentine and rather surprisingly rot- and moth-proofed wool and other materials made from natural fibres. The proofing chemicals used in this latter instance, lauroyl pentachlorophenol and dinitro alpha naphthalene respectively, did not give corrosive vapours but formic acid and acetic acids had been used in preliminary conditioning of the fabrics to bring the material to a state where it would more readily absorb the proofing material. This conditioning treatment was part of the processing proposed by the manufacturer and has been widely used.

Organic acid vapour corrosion is typically encountered in packaging, but other types of corrosive vapours are encountered earlier in the life of components. Vapours from hydrochloric acid pickling baths, paint and resin curing, chloro- and fluoro- organic solvent degreasing baths, plating solutions, coal- and oil-fired furnaces and internal combustion engines, are all corrosive, and all may be encountered where metals are being processed and protective treatments are being applied, conditions where metals are most susceptible to contamination and corrosion.

Contamination of metal surfaces by corrosive ions in the early stages of manufacture may induce corrosion immediately; but even when it does not, unless the metal is cleaned so that contaminants are removed, corrosion may occur even after the component has been painted or treated with temporary protective. This type of contamination is best controlled by choice of processing and good 'housekeeping' practices applied to work throughput, air extraction, and control of processing with particular attention to metal cleaning and subsequent storage.

CORROSIVE VAPOURS FROM DECOMPOSITION OF PLASTICS

Most organic polymers oxidise or depolymerise on ageing or heating to give volatile fragments. Hydrocarbon polymers give mostly carbon dioxide, carbon monoxide and water, which cause little corrosion, but under some conditions organic acids and aldehydes are produced. Polymers containing nitrogen, sulphur, bromine, chlorine and fluorine frequently decompose to give oxyacids of nitrogen, hydrogen sulphide, sulphur dioxide, and halogen acids, all of which are corrosive, with the last being especially aggressive to a wide range of metals.

The toxicity of fumes from outbreaks of fire in modern buildings, which inevitably contain quantities of synthetic polymers, is unhappily fully established as the major threat to life rather than the more spectacular heat and flames. The same vapours have often proved to be very corrosive, so much so that nowadays the major damage from small fires, especially if they involve electrical cables, is often corrosion of metals exposed to the fumes or the cost of guarding against the high risk of later corrosion. A similar type of failure is sometimes met in electrical equipment, where corrosion occurs in storage as a result of prior exposure to the decomposition products from organic materials derived from overheating during proving tests at the end of manufacture. Sometimes the overheated component will have been detected and replaced, repairing the immediate damage but not preventing the wider long-term effects. A typical instance of this type of corrosion has been reported by Turner (1964), who attributed failure of electronic equipment in a guided missile to corrosion during storage of the stainless-steel gears and other metal parts of a mechanical timer by the products of decomposition of PVC insulation which had been overheated in earlier testing.

Turner (1964) continued his investigation by evaluating the corrosion hazard from the decomposition products of each of a range of polymers used in electrical equipment as lacquer, potting and sleeving materials. Metal test specimens were exposed within sealed jars containing sufficient water to saturate the atmosphere. The organic materials to be tested were contained within the test jar in a nichrome boat which was first heated by an electric current to about 550°C for 5 minutes and then subjected to diurnal temperatures cycling between ambient temperatures and 70°C over a period of 7 days. As a result of a series of such tests on copper coupons coated with cadmium, zinc, tin, chromium, silver, gold and solder, and on bare coupons of copper and aluminium, Turner classified a series of organic materials in the following three categories:

Highly corrosive:	Neoprene, polyvinyl chloride, Teflon, polyvinylidene fluoride, polysulphide potting compound.
Slightly corrosive:	Nylon and Mylar insulations, insulating varnish, epoxide potting compound (Ciba Araldite 6060), silicone sleeving, diallylphthalate moulding compound.
Non-corrosive:	Polyurethane insulation, melamine laminate, polyurethane insulation and potting compound, nylon laminate, insulating varnish (Grade CB, 2 types), epoxide potting compound (except Araldite 6060 (see above)), silicone potting compound.

(The original paper gives detailed US military specifications and proprietary descriptions for the above materials.)

From Turner's results (1964) it can be seen that the worst effects can be attributed to halogens or sulphur; the presence of silicon or nitrogen does not necessarily give corrosive volatiles.

VAPOURS EVOLVED ON AGEING

Some polymers evolve corrosive volatiles on ageing, a process which is accelerated by higher temperatures but often occurs at well below those which give obvious and rapid degradation. The most corrosive vapours come from nitrocellulose and halogen-containing polymers.

Decomposition of nitrocellulose is accelerated by acids and oxidising agents; since the major products of decomposition, nitric oxide and dinitrogen tetroxide, are acidic and are also strong oxidising agents, the decomposition is autocatalytic. Once begun the reaction continues to accelerate so that stored nitrocellulose film for instance may occasionally ignite. Fortunately nitrocellulose is now little used as a plastic material although it is occasionally encountered, plasticised with naphtha, in special uses, and as a lacquer. It is a major constituent of many propellant and commercial explosive formulations, and although stabilizers are added to these compositions which ensure that they will not decompose in the normal range of ambient conditions, care must be exercised in ensuring that acids or alkalis are excluded.

REFERENCES

Andrew, J. F. Stringer, J. & Weighill, R. J. (1977) *Proc. 6th European Cong. on Metallic Corros. London. Soc. Chem. Ind.*, p. 639.

Angew, F. (1915) *Chem.* **28** 1, 72.

Arni, P. C., Cochrane, G. C. & Gray, J. D. (1965) *J. Appl. Chem. Lond.* **15,** 305 and **15,** 463.

Burns, R. M. & Freed, B. A. (1928) *J. Am. Inst. Elect. Engrs.* Aug. 579.
Cawthorne, R., Flavell, W. & Pinchin, F. J. (1969) *Br. Corros. J.* **4,** 35.
Cawthorne, R., Flavell, W. & Ross, N. C. (1966) *J. Appl. Chem. Lond.* **16,** 281.
Clarke, S. G. & Longhurst, E. E. (1961) *J. Appl. Chem. Lond.* **11,** 435.
Donovan, P. D. & Moynehan, T. M. (1965) *Corros. Sci.* **5,** 803.
Donovan, P. D. & Stringer, J. (1968) Unpublished report.
Donovan, P. D. & Stringer, J. (1971) *Br. Corr. J.* **6,** 132.
Knotkova-Cermakova, D. & Vlckova, Y. (1971) *Br. Corros. J.* **6,** 17.
Licari, J. J. & Browning, G. V. (1967) *Electronics* April 101.
Rance, V. E. & Cole, H. G. (1958) *Corrosion of Metals by Vapours from Organic Materials.* London, HMSO.
Schikorr, G. (1961) *Werkst. u Korrosion* **12** (1), 1.
Turner, I. T. (1964) *Materials Protection* Sept. 48.
Watson, R. (1789) Chem. Essays. London, p. 365.

7

Contact corrosion

Contact between metals and non-metals in wet or humid environments may accelerate corrosion of the metal. The primary cause is migration of corrosive ions from the non-metal to the surface of the metal, but the non-metal may also provide a source of water retained either by physical absorption or by deliquescent compounds within the material. The physical presence of the non-metal hinders drying of the contact area of the metal, and contaminants migrate into the crevice and accumulate there. Oxygen shielding may prevent repassivation of some metals such as aluminium and stainless steel and pitting may then occur. Once corrosion has begun availability of oxygen is sometimes a rate-determining factor; thus, for instance nails in wood frequently corrode rapidly on and around the head, but very slowly on the lower part of the shank in the area remote from the atmosphere.

The range of materials which can cause contact corrosion is wide but a limited number have a history of problems and are frequently used in conjunction with metals either in manufactured items or in handling, processing, or packaging of metals.

METALS AFFECTED

The lower organic acids are often encountered as causes of contact corrosion and they rapidly attack steel, zinc, cadmium, lead and magnesium, with effects on some other metals, as described earlier in Chapter 6 when discussing vapour corrosion.

Copper is especially susceptible to corrosion by certain plastics and temporary protectives since it reacts readily with long-chain organic acids giving bright green corrosion products. Its affinity for fatty acids is such that it will react with dioctyl or dibutyl phthalate ester plasticisers when in contact with PVC. or with the soaps in greases, or the acetyl groups in cellulose acetate. Copper also reacts with any remaining amine catalyst in poorly cured epoxide resins, the long-chain esters in organic proofing compounds such as lauroyl pentachlorophenate, and with sulphur or sulphides. Zinc has been reported by Randall (1970) to have given problems in contact with PVC in disrupting the polymer chain by acting as a catalyst in the breakdown of PVC plastisols.

Sulphates and chlorides are often present in natural materials and these cause corrosion of most metals. Corrosion of zinc, steel and aluminium may be accelerated by copper and mercury salts used in proofing many natural materials against moths and micro-organisms.

Sulphides from rubber, or hair, cause copper and silver to tarnish even in dry conditions. The silver sulphide whiskers which sometimes form, conduct electricity and are hazardous in electronic assemblies.

MATERIALS CAUSING CONTACT CORROSION

The materials that cause contact corrosion, the corrosive effects they give and the active ions which they contain that directly stimulate attack of metals are discussed in the following sections.

WOOD, PLYWOOD AND PAPER PRODUCTS

Acetic and formic acid vapours from wood frequently induce corrosion of nearby metals; these acids can equally cause corrosion of metals in contact with the wood and the effect is usually worse because the acid concentration is greater and the wood provides a ready source of water. But some woods, most notably oak, western red cedar and sweet chestnut also contain non-volatile constituents which are an additional corrosion hazard to metals with which they are in contact. The products responsible are polyhydroxy organic compounds loosely classified as tannins, although paradoxically certain tannin extracts in use as pretreatments have been shown to give a remarkable degree of protection to steel surfaces. Similar beneficial effects of tannins have been proposed to account for the preservation of old steel nails preserved in the vicinity of tanneries or associated with tanned leather products. A typical instance of corrosion from contact with wood reported by Evans (1951) involved corrosion of bronze statuary removed from the Fitzwilliam museum for safe keeping during the 1939–45 war. Ancient bronze artefacts packaged in wood shavings showed pitting attack which he attributed to acetic acid derived from the wood, which caused both contact and vapour corrosion.

Some timbers become potent causes of corrosion through the presence of marine salts introduced incidentally through the timber being floated in estuaries or carried as deck cargo, but occasionally salt is added deliberately to suppress cracking of the timber during drying. A noted instance of corrosion from this last cause reported by Farmer (1962) arose on the metal components of pianos exported from North America to Africa. The pianos were made from maple-wood which had been soaked in brine before drying, and in their country of origin they remained in dry temperate conditions for most of their life and showed few adverse effects, whereas the salt residues produced disastrous corrosion of metal components when a batch of these pianos was exported to a hot humid climate in Africa. Metal components out of contact with the wood were not corroded.

A widely used process for treating timber to prevent biodeterioration involves impregnation with an aqueous solution containing arsenates and chromates, sometimes with copper or mercury additions. After treatment the wood is often kiln dried so that acetic acid vapour is released, as described in Chapter 6, and the wood may then induce corrosion of nearby metals; in addition the copper or mercury rot-proofing salts are particularly corrosive to aluminium, and the copper ions may accelerate corrosion of aluminium, zinc or steel in wet conditions.

The corrosion of metals by wood is discussed in the Department of Industry Guide No. 2 (1979), which gives examples and guidance on relative acidities of wood. It also describes the preservation of wood, and the degradation of wood by metals.

Extensive experience has shown that paper used as a primary wrap for metal components may cause corrosion unless impurities are kept at a low level. The major corrosive impurities are sulphate and chloride. Sulphate is introduced during the manufacture of paper from wood using either the sulphate or sulphite method for separating the cellulose from lignins. Chlorides and sulphates may also come from the wash water. A typical example of corrosion of brass from contact with contaminated cardboard is shown in Fig. 7.1.

The particular instance illustrated in Fig. 7.1 resulted in the disposal of ninety million rounds of small arms ammunition and the cause was shown to be the high chloride content of 0.25%, calculated as the sodium salt, in the cardboard. To avoid this type of corrosion, sulphate and chloride levels determined by extraction with hot water must be respectively below 0.25% and 0.05%, calculated as sodium salts, and the pH of the extract must be in the range 5.5–7.5. Sometimes the low level of contaminants which is needed to prevent tarnishing may be impracticable or too costly and somewhat higher levels of contamination may be tolerated and some positive protection provided by impregnation of the paper with inhibitors (e.g. sodium benzoate, sodium chromate, sodium metasilicate, benzotriazole). Cotton (1970) has shown that tarnishing of brass from contact with paper can be delayed by prior treatment of the metal with benzotriazole. Chromate treatments of copper, brass, zinc and aluminium can also delay tarnishing from contact with paper containing impurities.

SOURCES OF CORROSIVE RESIDUES IN PLASTICS AND POLYMERS

Corrosive ions in plastics may be derived from side reactions during curing or processing, from impurities in the chemical intermediates used in the preparation of the polymer or from incomplete reactions in the final polymerisation. Serious problems may also arise through decomposition of

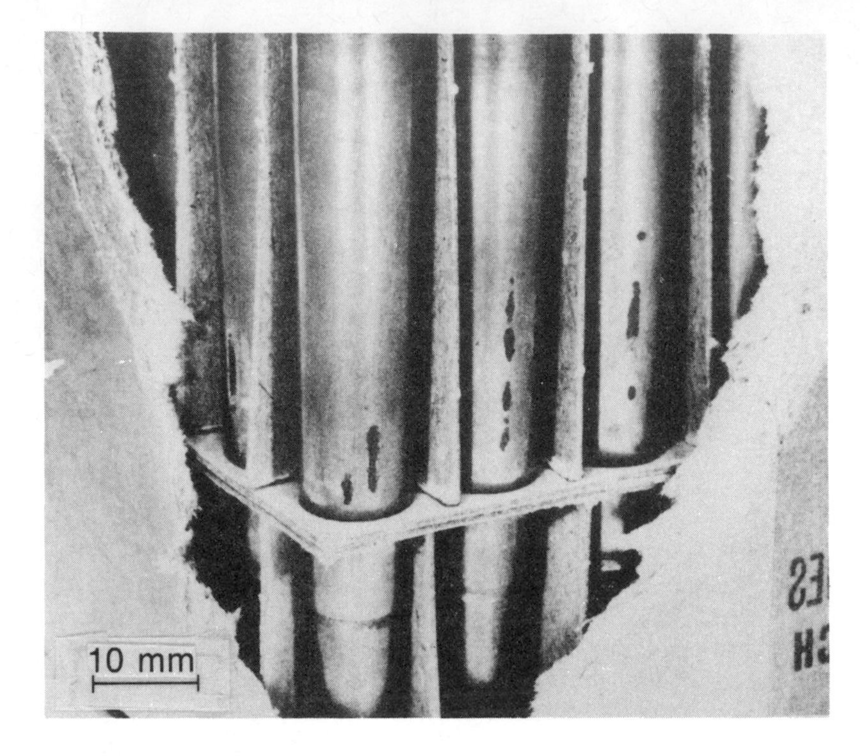

Fig. 7.1 — Corrosion of brass in contact with cardboard. (©Controller, Her Majesty's Stationery Office 1986.)

the polymer, from migration and decomposition of plasticisers, or from direct reaction between the polymer and a metal; the corrosion of copper through reaction with a cellulose acetate protective coating (see Figs. 13.2 and 13.3) is an example of this last type of reaction. Other examples are discussed in the following paragraphs.

POLYESTER RESINS

Polyester resins can cause both contact and vapour corrosion. Results of a test with three metals mounted in samples of a cold curing polyester are shown in Fig. 7.2. The attack on the wires in Fig. 7.2 is greatest where the metal wires protrude through the resin–air surface; within the resin the metals remained uncorroded. Because of this risk polyesters cannot be used for potting electrical assemblies and are at a disadvantage as materials of construction for packages and other structures which may bring them into contact with metals.

The most corrosive polyesters are the cold-cured formulations for which the more reactive peroxide catalysts, such as methyl ethyl ketone hydroperoxide are usually used. The short-chain acid fragments from these aliphatic

Fig. 7.2 — Corrosion of metal wires by polyester after four weeks at 30°C and 100% RH. (From left to right: iron; copper; cadmium.) (©Controller, Her Majesty's Stationery Office 1986.)

peroxides are generally more corrosive than those such as benzoic acid from aromatic peroxides which are used in hot-curing formulations. Hot curing is also beneficial as it encourages the escape of any volatile fragments which are formed. Chlorinated reactants, such as HET-acid (chlorendic acid), are often used to give fire retardant grades and these may present an even higher corrosion risk.

Ethylene glycol is widely used in polyester formulations but gives a relatively high level of volatile acids through the breakdown of ether peroxides which are formed by reaction between the ether group in the glycol and the peroxide catalyst. Less corrosive resins are formed if a glycol such as propylene glycol without ether linkages is used; if benzoyl peroxide is used as catalyst with *NN*-dimethyl aniline as accelerator, and with an addition of 1% sodium bicarbonate to the final mix, then the resulting resins are almost non-corrosive. Further improvement may be obtained by post-curing at 120°C. Andrew, Stringer & Weighill (1977) have demonstrated that glass fibre reinforced packages based on this formulation are not corrosive in contact with steel, zinc, cadmium, magnesium alloy or brass even in hot moist conditions. But these workers emphasise that care must be exercised in selecting glass mat reinforcement and mould release agents to ensure that they do not contain PVA.

HALOGEN-CONTAINING POLYMERS

PVC is the least stable of all commercial polymers but one of the most widely used. It begins to decompose at 60–70°C releasing hydrogen chloride which is very corrosive to a wide range of metals. Fluoro-polymers are much more stable, only decomposing when the polymer reaches about 350°C, but all the halogen polymers will cause intense corrosion once their threshold for decomposition is reached. Copper corrodes in contact with PVC plasticised with phthalate esters, to give bright green corrosion products. Sometimes this affects the polymer which begins to decompose causing even more serious corrosion. Zinc may catalyse the decomposition of PVC although the effect may be countered by the use of lead carbonate stabilisers (Randall, 1970). Migration of solutions and plasticisers along PVC sleeving is a constant cause of problems with electronic equipment in wet or humid conditions. The migrating moisture collects and transports impurities which may emerge at areas well away from the source of the water by-passing seals and reaching vulnerable components. Plasticised PVC may be a cause of serious deterioration in contact with other polymers; ester plasticiser may soften paints and potting resins. A particular problem encountered by the writer involved the complete corrosion of tinned copper wiring within PVC sleeving. The outside of the PVC was in contact with nitrocellulose plasticised with naphtha. The corrosion product contained nitrates, chlorides and organic copper salts.

PHENOLICS AND FORMALDEHYDE RESINS

Undercuring and the use of wood-flour fillers are the causes of most corrosion problems reported in service with cross-linked phenolic/formaldehyde resins produced by the single-stage process, but two-stage curing resins are prolific sources of ammonia and formaldehyde.

EPOXIDES

Epoxides are unreactive to most metals but they do contain a small proportion of chloride from the epichlorhydrin from which the epoxide is made, and this occasionally causes slight corrosion. Cold-cured compositions are often cured with amines and these may tarnish copper alloys.

RUBBERS

Natural rubbers oxidise to organic acids and when vulcanised contain reactive sulphur. Copper in contact with vulcanised rubber may react with the sulphur and in doing so degrade the rubber. Many synthetic cold-curing rubbers are corrosive; some contain acid or ester solvents and may be very corrosive, while others contain sulphides and tarnish copper. Many are cured with peroxide catalysts which give traces of organic acids. A particularly corrosive silicone sealant encountered by the writer had been selected for a critical application because of the reputation silicones possess for inertness. The sealant contained acetic acid as solvent which was retained for several weeks and the sealant remained very corrosive. Rubbers are a frequent cause of corrosion and should be selected with caution. Chlororubbers are the least reactive at ambient temperatures but may be corrosive when heated above about 75°C.

RESINS WITH INTEGRATED CIRCUITS

The effects of resins on the performance of silicon monolithic microcircuits is an area of great current interest since resin encapsulation gives robust products and great financial saving over sealed metal packages. The reactions involve chemisorption and the formation of ordered dipoles by migration of ions into 'holes' in the outer layer of the devices; unacceptable changes in performance happen before any conventional changes occur. Licari & Browning (1967) have identified possible contaminants in resins which could have adverse effects on microdiodes. The performance of epoxides, phenolics, silicones and fluorocarbons are reviewed; phenolics and alkyds were both shown to degrade the performance of microdiodes and ways of alleviating the effects are discussed. The present aim of the electronics industry to produce silicon wafers with 1 million gates and conductors 1 μm wide on devices 2 mm in diameter and 750 μm thick is requiring new concepts of chemical purity and materials compatibility and is an area of advance giving new insights into the reactivity of surfaces and the effects of impurities.

CEMENT AND CONCRETE

Concretes made from Portland cement consist of an inert aggregate contained in a matrix of hydrated silicates and aluminates with a small proportion of lime and about 4% calcium sulphate. The sulphate is liable to cause corrosion of many metals but the lime maintains alkaline conditions which reduce the probability of corrosion of most metals although aluminium may be attacked; on ageing the lime reacts slowly but progressively with atmospheric carbon dioxide and its beneficial effect is reduced. Calcium chloride is sometimes added to the concrete mix to accelerate setting and to prevent freezing during cold weather. Concretes made with such additions, which are typically up to 4% of the weight of cement used, are likely to induce corrosion of metals in contact with the concrete. Occasionally active ingredients in the aggregate may cause corrosion; boiler slag, gypsum residues and clinker for instance have reputations for causing corrosion because of their high sulphur contents.

The fine capillary network within concretes provides a ready path for the migration of moisture which may carry soluble salts, and the abrasive surface may damage or remove protective coatings. These two effects may cause corrosion in some circumstances but again the alkaline environment reduces the risk.

A composition based on a magnesium oxychloride matrix, usually with an aggregate of limestone chippings, is sometimes used to provide a flooring with a high standard of surface finish. The high chloride content of this material ensures that it is highly corrosive to most metals. In one particular instance a set of magnesium castings placed on a floor of this type was seriously corroded within a few weeks.

Wall and ceiling plasters, and many plasterboards are based on calcium sulphate and will cause corrosion of most metals in humid conditions.

The chemical ingredients are only part of the corrosive risk from floors; additional contaminants are introduced through spillages, de-icing salts, settlement of dust and soot and the absorption of vapours. The hard gritty surface of concrete floors readily abrades away protectives from metals in contact, and temperature differentials, osmotic effects, washing and spillages all tend to result in frequent wetting of surfaces in contact. Even surfaces out of contact with the floor but close to it become contaminated by dust, splashing and solution creep through surface tension so that corrosion spreads beyond the immediate area of contact. Metals should therefore be stored above floor level.

DRYING-OIL PRODUCTS

Most air-drying paints are based on drying-oils, and these and other drying-oil based materials such as linseed oil putties, may cause corrosion. As these materials when applied to metal surfaces reduce access of oxygen and water and also have an inhibitive content there is a measure of competition between their corrosive and protective effects. With zinc the aggressive effect predominates so that unless special formulations are used, under moist conditions the zinc corrodes and the paint or putty flakes away. Steel in contact with drying-oil formulations in damp conditions may sometimes similarly corrode, especially beneath putty, which is low in inhibitors.

Drying-oils cure through a free-radical cross-linking reaction which is initiated by absorbed oxygen acting on the conjugated double bonds of the drying-oil but some side reactions occur, similar to those referred to earlier with polyesters, giving formic and other organic acids (see also Chapter 6). Certain basic pigments such as calcium plumbate when added to paints react with these short-chain organic acids releasing plumbate ions which are effective inhibitors so that the net effect is to remove the aggressive ions and strengthen the inhibitive effect. Priming paints with such additives are usually compatible with zinc surfaces, although in warm humid atmospheres, especially in enclosures, adhesion failures may occur.

A range of synthetic resins are available and should be considered where drying-oils are likely to give problems. In particular, epoxides and polyurethanes give high quality inert paints which are compatible with most metals and plastics.

REFERENCES

Andrew, J. F., Stringer, J. & Weighill, R. J. (1977) *Proc. 6th European Cong. on Metallic Corros.* Soc. Chem. Ind., p. 639.

Cotton, J. B. (1970) *Proc. Conf. on Protection of Metals in Storage and Transit.* Lond. Brintex Exhibitions.

Evans, U. R. (1951) *Chem. and Indust.* 710.

Farmer, R. H. (1962) *Wood* Nov. 444.

Licari, J. J. & Browning, G. V. (1967) *Electronics* April 17, 101.

Randall, F. G. (1970) *Br. Corros. J.* **5**, 144.

8

Corrosion by residues from processing

Many of the engineering metals when contaminated with certain active impurities will corrode if they are exposed at high humidities even in a clean atmosphere. The corrosive contaminants most frequently encountered are chlorides, sulphates, sulphides, nitrates and organic acids. A contaminated surface will remain uncorroded while it is stored in a dry atmosphere but will corrode once the humidity rises above the critical level. The same metal exposed to high humidity without contamination remains unaffected.

Careful packaging and overcoatings of paint or temporary protectives may not suppress the effects of corrosive contaminants already on the surface, since oxygen and water readily diffuse through organic coatings and react with the metal under the influence of the active ions, and damaging corrosion may blister the coating and spread. Inhibitors may suppress the reaction for a time but overall the protection afforded by the coating is diminished.

Metals may become contaminated during manufacture and processing by residues from acid pickling, electroplating, chemical conversion processes, heat treatment baths, degreasing fluids, soldering and brazing fluxes, machining oils, crack detection fluids, lubricants used in metal rolling and drawing, sweat and heavily contaminated atmospheres such as those in plating shops and in or near furnaces or stoving ovens.

Contamination is best avoided or removed by: selecting processes which are free of the more corrosive ions; ensuring that adequate cleaning processes are introduced; adhering to good design practice by avoiding cracks, crevices and cavities which are liable to trap residues and are difficult to clean and protect; ensuring that metals are clean before heat treatments so that trapped organic halogen degreasing and crack-detecting fluids, fluxes and drawing lubricants are not degraded and react with the metal surfaces; keeping cleaned items in segregated areas away from chemical treatment, workshop, painting and heat-treatment areas and ensuring that clean metal surfaces are not handled with bare hands. Residues are especially difficult to remove from items produced by certain processes such as casting and sintering in which a degree of porosity is almost inevitable; treatment of such items in solutions containing aggressive ions should be avoided, dry treatments being preferred where possible.

PICKLING SOLUTIONS

Hydrochloric, sulphuric and phosphoric acid are widely used in removing oxide scale from metal surfaces. The chloride residues from hydrochloric acid are the most corrosive contaminant and as this acid is volatile, not only treated metals are at risk from corrosion by retained residues but so are any nearby surfaces. However hydrochloric acid is attractive for metal cleaning because treatment times are short and heating is not required. For processing large quantities of heavily scaled metal these advantages may outweigh the risk of corrosion. If metals are pickled in hydrochloric acid they must be well washed immediately after treatment and removed to another area.

Sulphuric acid is less volatile and the residues are generally less corrosive than those of hydrochloric acid, although it is usually necessary to heat the bath to about 60°C to ensure an adequate rate of reaction with steel. Fig. 8.1 shows a component taken from storage after two years which clearly shows drainage marks and heavy staining developed from residues of a pickling treatment in sulphuric acid at the end of production; the item would have been subjected to an intense examination before packaging to ensure that the surface was clean and free of visible defects.

Plain phosphoric pickling acid has the considerable advantage that traces cause little subsequent corrosion to the majority of metals but it is relatively expensive and for pickling steel needs to be heated to about 85°C; the greater initial cost and the need for heating discourage the use of this otherwise admirable acid.

Electroplating solutions often contain aggressive ions which may be retained in the same way as pickling solutions. The effects of crevices and pits are made worse by the plated coatings which narrow or seal the mouths of crevices and pores, and thus trap plating solutions which exude later to produce unsightly stains; an effect often referred to as ‘spotting out’.

CONVERSION COATINGS

Conversion coatings are reviewed in Chapter 11. Many are based on solutions which contain aggressive ions and give rise to carry-over of corrosive residues. Phosphating of steel in accelerated baths and anodizing

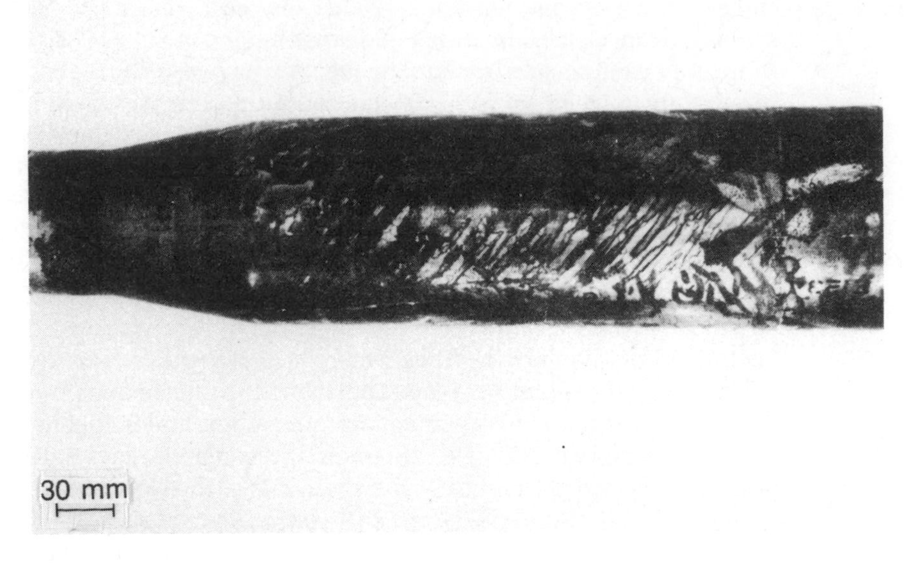

Fig. 8.1 — Staining from acid pickling residues on brass after two years of storage. (©Controller, Her Majesty's Stationery Office 1986.)

of aluminium are two processes which illustrate the risks involved; both processes are widely used and the risk of corrosion from retained residues has been an important factor in the development of commercial processes.

The earliest phosphating baths consisted of a solution of iron, zinc or manganese dihydrogen phosphate ($Fe(H_2PO_4)_2$ + $Mn(H_2PO_4)_2$) with a slight excess of phosphoric acid. Steel immersed in the bath is attacked by the acid with evolution of hydrogen and crystals of metal monohydrogen phosphate ($FeHPO_4$) are deposited on the metal surface until the surface is completely covered. Treatment takes about 40 minutes at 95°C and gives heavy coatings. Reduced treatment times are obtained by using accelerated baths. These contain additions of inorganic nitrates, or an organic nitro compound, which provide more efficient cathodic reactions than hydrogen release. The treatment time is reduced to about 15 minutes and the working temperature is lowered to 75°C. The inorganic nitrate is a corrosive hazard if it is subsequently retained on the metal surface, and for this reason an organic nitro compound, usually nitroguanidine, is often preferred since, although it is more costly and slightly less efficient, its residues are not aggressive.

Prolonged processing in treatment baths is difficult to incorporate into modern high-throughput production lines and even more reactive phosphating systems have been developed based on stronger oxidising agents, often chlorates or perchlorates, as the accelerator. These are applied by spray and produce a complete coating within about a minute. Coatings are very thin, which is an advantage in coating thin sheet such as that used in car bodies since paint on thicker phosphate coatings tends to crack on flexing. Any retained chlorides from these spray treatments are very corrosive and thorough rinsing treatments are needed. A final rinse in dilute chromic acid solution is often effective in countering the risk from retained residues; traces of chromate act as an inhibitor and combat some of the effects of the aggressive residues. In practice it has been found that the strengths of such chromic acid solutions should not exceed 0.05% since greater concentrations have been shown to be a possible cause of dermatitis to those who later handle the treated components. There are, however, continuing moves to a greater stringency in the regulations covering the discharge of effluents and the use of toxic materials, which increasingly inhibit the use of chromic acid solutions in industrial processing.

Anodizing of aluminium has parallels to the phosphating of steel. The earliest protective anodizing treatment, developed by Bengough, was based on treatment in chromic acid solutions. Although by later standards the films were relatively thin, they gave a valuable degree of protection and any retained residues were not harmful. Later a sulphuric acid treatment was developed which gives thicker coatings that can be coloured more effectively than the chromic acid coatings. The thicker coatings however are more likely to bridge the openings of small cracks and pores so that there is a risk of trapping residues, but since a final part of the process is a prolonged sealing treatment in hot water, the risk is low unless articles of unsuitable geometry or of poor finish, or with cracks or pores, are treated. An extra measure of safety is sometimes provided by a final sealing in a hot chromate solution. One of the few remaining uses of the original Bengough chromic acid anodizing process is to reveal any cracks or surface imperfections in aluminium castings. Once anodized the items are given a rapid rinse and then allowed to stand when coloured chromate exuding from cracks and pores renders them visible. A second remaining use of the Bengough process is in treating surfaces which are to be in contact with explosives or propellants where the risk from any reaction with sulphuric acid residues is unacceptable; the high visibility of any chromic acid makes visual inspection for retained residues effective.

HEAT TREATMENT, DEGREASING AND CRACK DETECTION

Heat treatment baths based on molten salts are an obvious source of corrosive residues. Chlorides again should be avoided if possible. The same general strictures referred to with pickling treatments are appropriate — thorough washing and immediate removal from the vicinity of the treatment area.

The organic halogens used in degreasing and crack detection are normally stabilised by additions of inhibitors but in contact with some metals they may decompose to give very corrosive residues; the conditions under which this is liable to occur and appropriate precautions are discussed in Chapter 10. Vapours from these processes should be well vented and metals should be stored well away. A less widely appreciated risk arises from the entrapment of traces of these organic halogens in crevices and within box sections of components which are then subjected to heat treatments where the trapped residues may decompose to give corrosive halogen acids. Fig. 8.2 shows a typical example, the inside of a box section of FV520S heat-

Fig. 8.2 — FV520S steel corroded by residues from crack detection fluid: three months after manufacture. (©Controller, Her Majesty's Stationery Office 1986.)

treated stainless steel which in common with other stainless steels is rusted by severe chloride environments. After welding, the unit shown in Fig. 8.2 was immersed in a bath of crack-detecting fluid based on an organic halogen solvent. Hardening followed and involved heat treatment at 475°C. Any solvent remaining within the box section would have decomposed at this temperature and after storage in an outdoor environment the chloride residues had caused extensive general rusting and some pitting on the internal surfaces of a steel which is normally considered to resist rusting indefinitely; certainly the external surfaces were unaffected.

SOLDERING

For effective soldering, surfaces must be cleaned, usually by first degreasing and then, if necessary, by machining or abrading the surface. Phosphoric acid pickling is preferred for cleaning. If the more reactive hydrochloric or sulphuric acid is used, residues must be thoroughly washed away before soldering.

Fluxes used in soldering keep the surfaces free from oxides so that the solder can effectively wet the surfaces being bonded. Traditional fluxes have been based on chlorides — usually zinc chloride ('killed spirit': sometimes made by adding granulated zinc to dilute hydrochloric acid) or ammonium chloride. Experience has shown that even with diligent cleaning the risk of corrosion from residues of chloride fluxes is too high to be tolerated and less corrosive fluxes have been developed and are to be preferred even though their use may require more precise control.

Forty per cent phosphoric acid is an effective flux for steel and copper alloys but residues are not entirely free from corrosive effects and it is not a fully 'safe' flux although it is widely and successfully used especially for stainless-steel joints.

A range of resin fluxes which are not corrosive to metals is now available. They are based on resin (or rosin), a natural product obtained from pine trees which consists largely of abietic acid. It can be applied as a powder obtained by crushing the resin, or better brushed on as an alcoholic solution. More tolerant resin fluxes sometimes called activated resin are used; these contain additions of activators such as aniline hydrochloride or lactic acid which volatilise leaving little residue, but both aniline hydrochloride and lactic acid may cause some staining, especially of copper alloys. Etheridge (1980) has provided a useful review of tests for 'non-corrosive' soldering fluxes; the conference proceedings in which Etheridge's paper is published contain other valuable papers on aspects of the use and formulation of non-corrosive fluxes.

Recent attention in the electronics industry has been turned from resin fluxes, which require organic solvents and present effluent problems, to water-soluble products such as polyethylene glycol or triethanolamine. An efficient flux of this type is claimed by Schuessler (1980) to be non-ionic, non-acidic and free of organic solvents.

MACHINING OILS

Oil emulsions are widely used to cool and lubricate metal surfaces during rolling, grinding and cutting. Films from these processes often remain on the metal surface where they are usually harmless and indeed offer some protection, but occasionally they cause unsightly staining. The oils are usually fed from a central reservoir to which they are returned in a recirculating system. In the static tanks micro-organisms may grow in the aqueous phase, finding nutrients in the oil and in any contaminants, and they excrete a range of products including organic acids. The micro-organisms themselves may build up as slimes and sediments blocking circulation and degrading the properties of the oil. In a review of this topic, Hill (1968) asserts that there is good evidence that mottled markings on rolled steel sheet and allied problems on galvanized and tinned steel sheet have their origins in the organisms present in the rolling mills. He recommends a 'red-spot' test for suspected oils in which a drop of oil or emulsion is incubated on a special bacterial growth culture for 16 hours when a red colour develops if bacteria are present. Smells and blocked filters are a sure sign of infection. Biocides and good housekeeping — especially prevention of contamination of the oil by potential nutrients, are needed to counter the risk.

HANDLING RESIDUES

Sweat deposited on metals from handling is a frequent source of subsequent staining and corrosion. Fig. 8.3 shows a typical example: a brass case which

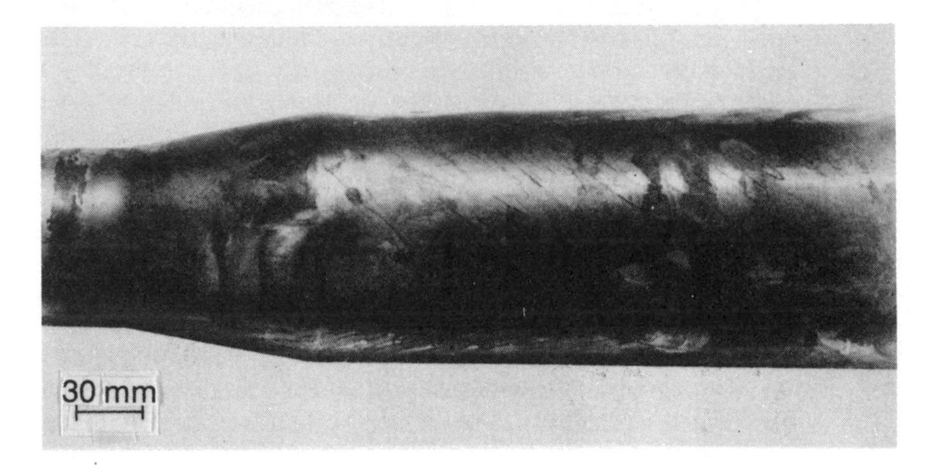

Fig. 8.3 — Hand prints on brass developed after two years of storage. (©Controller, Her Majesty's Stationery Office 1986.)

was found to be heavily stained after two years of storage. The pattern of the original hand prints responsible for the stains is clearly visible.

Close examination of a freshly sweat-stained surface shows the deposit to consist of a grease phase and droplets of an aqueous phase. Both are rich in organic acids with butyric acid being the major constitutent; the aqueous phase also contains a high concentration of inorganic salts. Studies on the relative corrosivity of sweat from different individuals have indicated a wide variation in pH and 'corrosiveness', and systems for excluding the most corrosive hands have been suggested and may have some limited success but leave the fundamental physiological phenomenon uncontrolled and the basic problem unsolved. Both high temperatures and high humidity increase the extent of sweating, which usually increases as the day proceeds, stimulated by the activity of work and by rising temperatures. In some regions the practice of eating fruit, particularly citrus fruit, during meal breaks in hot weather exacerbates the effects both by increasing the quantity, and in extending the variety of contaminants present so that the effects are pronounced and varied.

Gloves should be worn for handling clean metals but the difficulties of ensuring that this precaution is enforced should not be underestimated. Gloves increase the extent of sweating and in hot conditions become uncomfortable and the temptation to remove them for delicate tasks is difficult to resist; any objects touched under such conditions will be exceptionally heavily contaminated. A preferred alternative, where possible, is to arrange that all handling is carried out with mechanical aids using racks, hooks or trays. Handling in final inspection is an additional hazard which may be minimized by using racks or jigs, or better, avoided by automatic monitoring. Holders must of course be kept clean or washed if they become contaminated. Their progress through the system must be properly controlled as racks carelessly transferred from contaminated workshops to clean areas may be a potent source of residues. An allied problem arises from the use of locally 'designed' transit packs between stages of manufacture or inspection, or on transfer to final testing and packaging areas. These transit packs are often unspecified, uncontrolled and unknown to the designer of the process, and sometimes contain all the worst features of bad packaging so that surfaces which may have been maintained in a clean condition become heavily contaminated just before they are finally packaged.

HEAVILY CONTAMINATED ATMOSPHERES

Metallic components often spend days, weeks, even months passing through the various stages of fabrication, processing, assembly inspection and awaiting final despatch. Inevitably, items between these stages have to be stored near processes which may include furnaces, rolling mills, heat treatment baths, acid pickling solutions, machining operations in which aerosols of oil are generated, plating baths, pretreatment processes, degreasing solutions containing hot solvents, grit blasting, metal spraying and painting. Good practice requires that metal components be stored in separate areas with independent ventilation systems but all too often many or all of the processes of production are carried out in one large building with few subdivisions.

Items in inter-stage storage are sometimes held in racks or shelving within the same large enclosed space sharing its air circulation with the processing areas. The circulating air, which inevitably becomes contaminated with aerosols, vapours and dust, reaches metal surfaces and contaminants may be absorbed or condense in the upper spaces of the building and drip, or dry out and fall as solid debris, onto the metal surfaces below. Overwraps and covers over work in progress prevent the worst of these effects but storage in separate areas is also advisable.

Air for storage areas should be drawn from a clean area outside and filtered, with the relative humidity of the air inside being controlled by heating and, as necessary, with desiccants. Even here difficulties can arise; Scott & Skerrey (1970), for instance, describe how a heated building used for the storage of aluminium became contaminated because the combustion products from the heating furnaces were released inside the building. They estimated that in the four months before the situation was rectified 18 kg of sulphur dioxide (and a large quantity of water) had been discharged into the building.

Contaminated metals may corrode before they are packaged, or as frequently happens, incipient corrosion interferes with the next process, particularly when this is plating, or some types of decorative painting. The consequent defects in the form of poor adhesion, excessive porosity or thinned areas of coatings become apparent after a period of storage. If the contaminant gives none of these problems immediately it will remain and staining or more drastic corrosion may occur later when high humidities are encountered.

REFERENCES

Etheridge, A. J. (1980) *Proc. Internepcon UK 80*, Brighton, p. 80.

Hill, E. C. (1968) Walters, A. H. & Elphick, J. J. (Eds) *Biodeterioration of Materials*. Elsevier Publ. Co., p. 381.

Schuessler, P. (1980) *Proc. Internepcon UK 80*, Brighton, p. 68.

Scott, D. J. & Skerrey, E. W. (1970) *Proc. Conf. Protection of Metal in Storage and Transit,* London, Brintex Exhibitions, p. 45.

9

Corrosion and deterioration by biological processes

Corrosion of metals in storage may be accelerated by biological activity, usually as a secondary effect resulting from primary damage caused to protective coatings or packaging materials, or from the products of biological activity. The only established reaction where biological activity plays a direct role in corrosion is the cathodic depolarisation of hydrogen ions in the corrosion of steel caused by the enzyme hydrogenase which is produced by the *desulphovibrio* micro-organism. This micro-organism accelerates the corrosion of iron in anaerobic (oxygen-free) conditions, typically on buried structures. Other reactions such as the conversion of sulphates to sulphides are also used by this micro-organism in a cycle of reactions that leads to the corrosion of buried steel (see for instance Tiller (1982)). Occasionally and conversely the presence of a corroding metal may suppress biological activity, as copper was shown to do on the hulls of wooden ships so that copper-bottomed has become synonymous with reliability.

DAMAGING ORGANISMS

The organisms responsible for the corrosion of metals and deterioration of other materials may be considered in the following groups:

Micro-organisms. Minute plants or animals which individually may only be observed under a microscope although collectively they may be readily visible. Subdivisions include: fungi and moulds which are multicellular organisms lacking chlorophyll and are conveniently considered in this class although the primitive cells are loosely associated and grow together as fibrils giving various wool-, felt-, hair- or thread-like structures. Bacteria, yeasts and algae are unicellular organisms; algae contain chlorophyll and require sunlight for growth.

Insects. Six-legged invertebrates which develop from eggs and pass through moults often involving larval (grub, maggot or caterpillar) and pupal (chrysalis) stages. Damaging varieties include termites, wool moths, wood wasps and wood-boring beetles.
Mites. Very small primitive, eight-legged invertebrates.
Rodents. Gnawing animals — rats, mice and squirrels.

To subsist and multiply, living organisms require water and food. The smaller organisms may obtain water from the air, or from materials in equilibrium with air, at relative humidities above about 70%. There are numerous varieties of small organisms which may cause degradation; each is adapted to specific types of food and living conditions although mutations are readily generated which can accomodate new conditions. The micro-organisms are most prolific at temperatures within the range 15–35°C and at relative humidities above 85%. They usually find food from among natural products such as wood, paper, wool, cotton, jute, petroleum products, natural fats, oils and resins, animal- and vegetable-based adhesives, and certain esters, organic acids and amines. Purely synthetic polymers are much less readily attacked but they are not immune; examination of buried samples has revealed that even polythene film can be penetrated by biological activity. The plasticisers present in some synthetic materials such as PVC provide a source of food for some organisms with damaging effects to the plastic. Films of dust, dirt and handling residues encourage the growth of micro-organisms through the extra variety of diet they offer. Insects thrive in similar but more restricted habitats to micro-organisms; they need natural materials as food and most are inactive outside the temperature range of 10–40°C.

ATTACK BY LIVING ORGANISMS AND THEIR CONTROL

Fungi, bacteria, yeasts and algae are ubiquitous in their occurrence and even if not already present in an enclosure they will penetrate small leakage paths, even through pin holes in metal foil. Varieties are encountered thriving at temperatures from −10 to 100°C, but the vast majority prosper at temperatures from 15 to 35°C. Most require oxygen but some are adapted to live without it and others only thrive in its absence.

Materials attacked by micro-organisms may become discoloured or stained, lose strength or other physical properties, smell, and may disintegrate. Growths of micro-organisms provide food for mites and insects, the bodies and detritus from which may cause corrosion, and their activities may damage packaging and protectives. Micro-organisms can destroy packaging, barrier materials and protective coatings on metals; convert water/oil emulsions into foul-smelling corrosive mixtures; produce fibrous growths in water–fuel interfaces which block fuel lines; cause short circuits on electronic components; generate organic acids and amines which may corrode nearby metals; and degrade plastics and corrode buried metal structures.

Miller (1980) in an informative discussion of the principles of microbial corrosion in relation to metabolic and energy considerations identifies three main mechanisms:

— absorption of nutrients (including oxygen) by microbial growths adhering to metal surfaces;
— utilisation of various substances that produce deleterious changes in the physical or chemical properties of the environment, e.g. breaking up or forming emulsions, or degrading antioxidants;
— producing damaging end-products or by-products of growth, particularly: (a) surfactants, (b) sulphuric acid, and (c) organic acids.

Insects may damage or even destroy buildings, packages and packaged materials through eating, boring, and by the corrosive action of their waste products and decomposition of their bodies. Insects enter enclosures as eggs with the contents, or as freshly hatched larvae through holes and crevices. The bore-holes which often betray their activities are usually the final exit of larger larvae or mature insects. Most natural organic products are attacked by insects, with wood, wool and paper products being favourite targets; they are most active in humid tropical conditions. Mites usually feed on fungi and can be active at relative humidities as low as 50%; their presence attracts insects and their waste products may cause corrosion.

The brown, Norwegian, or common rat, the black or ship's rat and the house mouse are the major rodent pests. All breed prolifically, are adaptable in their habits and feeding, and damaging in their activities. They are diligent and determined in gnawing harder substances such as wood to condition their teeth and shred softer materials for nesting. They will travel to obtain the food and water essential to their existence. The conditions of storage areas are often well suited to their needs, especially in mixed storage when foodstuffs are also present, and they are a major cause of deterioration. Squirrels adapt to some storage conditions and can be a similar nuisance.

The methods for controlling biological attack share a common approach, and are:

— maintain dry conditions, below 60% RH if possible. Dry materials before introducing them into a store or package and keep sources of free water covered;
— sterilise and fumigate storage areas and containers. Keep returned, or otherwise suspect, containers in segregated areas until they have been cleaned, and preferably sterilised;
— protect susceptible materials by resistant wraps, screens and metal reinforcements;
— use resistant materials when possible. Thus synthetic plastics are more resistant than wood and paper, and hard woods, chlorobutyl rubber and synthetic resins are respectively more resistant than soft woods, natural rubber or resins based on natural products. Fillers and plasticisers in otherwise inert resins may be attacked by micro-organisms;

— use chemical impregnations and additives to increase resistance to attack;
— ensure by good practice and design that pests are denied access to storage areas and that, if they do enter, ready harbours and breeding places are not available.

The principles involved in maintaining dry environments are discussed in Chapter 14, and these are equally applicable to the control of micro-organisms and insects. To discourage rodents, all sources of free water in drains and supply tanks should be covered, with no gaps wider than 6 mm; care should be exercised in preventing taps from dripping, and any leaks or ingress of ground water should be attended to immediately; food should not be accessible.

Within storage areas any signs of pests should be met with removal of infested and suspect materials, cleaning, preferably by vacuum treatment, and with dusting or spraying with insecticides, or by fumigation. Buildings should be secured to prevent ingress of rodents with a clear space surrounding them. Precautions must be extended to vehicles carrying goods; they should be kept closed, never left in infected areas, and thoroughly cleaned at regular intervals. In buildings, enclosures, cavities and ducting should be well covered and regularly inspected. Crevices and rough surfaces should be avoided since they may harbour insects and offer pathways to rodents which can readily scale vertical surfaces it they are sufficiently rough. Coving should be preferred to angular joints. Edges and corners of any wooden doors or containers should be covered with metal strip to deter rodents and when the hazard is great, wooden surfaces may have to be covered with metal gauze. Heavy rodent infestation needs expert attention but permanent baits may control lesser levels.

A range of chemical treatments is available to render most materials resistant to microbes and insects. These are usually effective when applied in sufficient quantity but may degrade or leach out if exposed for long periods to the sun or to heavy condensation. The need for compatibility with adjacent materials may limit the range of chemicals that can be used. They must not of course degrade the material being treated and the method of treatment must give a sufficient loading; for best effect deep penetration is needed. Cost is also important. Widely used preservatives include creosote (a cheap and general-purpose wood preservative but incompatible with many metals and organic materials); copper, zinc and mercury salts (mercuric salts have sometimes been used on wood but they are incompatible with some metals, especially aluminium, and their use is limited in some countries); di-nitro alpha naphthol (DAN, used in moth-proofing wool); lauroyl pentachlorophenate (LPCP, for rot-proofing fabrics: not compatible with copper); pentachlorophenol (fungicide and effective against termites); boron compounds; fluorides; chromates; arsenates; alpha-chloronaphthalene; chlordane; dieldrin; gamma-benzene hexachloride; and tributyl tin oxide.

WOOD AND WOOD PRODUCTS

Wood is susceptible to attack by micro-organisms, suffering brown rot, white rot, soft rot, dry rot and blue stain. It provides food and home to termites, beetles and wood wasps, and is the preferred gnawing material for rodents. Soft woods are very susceptible to all these problems while western red cedar and some of the hard woods, particularly oak, are relatively resistant to attack. Micro-organisms and insects will not thrive on wood if the moisture content is kept below 20%. Preservative treatments are advisable in most environments and are essential in wet tropical and semi-tropical environments. Boards, papers, wood-flour filling in resins, starch and cellulose pastes and other wood products are susceptible to attack by some of the same pests as wood and require similar protection.

BS 1133 (1967), although in need of revision, provides much useful information on the use and preservation of wood and other natural products. BS 1282 classifies wood preservatives and their application, and BS 3452 gives details of the compositions and application of water-based copper/chrome wood preservatives. Advisory leaflets on all aspects of wood and its use are available from the Forest Products Princes Risborough Laboratory, and in HMSO publications.

WOOL AND NATURAL FIBRE FABRICS

Wool products are prone to attack by moths, and cottons by fungi. All require treatment with biocides and fungicides. Some treatments involve prior conditioning with formic or acetic acid to ensure adequate take-up of the active species. The volatile acid in the treated fabric may be a vapour corrosion hazard to nearby metals. Lauroyl pentachlorophenate is widely used for rot-proofing cotton fabrics but in contact with copper alloys it decomposes to give a build-up of copper salts on the metal.

PAINTS

Water-thinned paints which often contain cellulose, proteins, PVA and emulsifying and stabilising additives are susceptible to microbiological degradation during storage. Fungicidal and bactericidal additives are therefore essential ingredients.

Oleoresinous and resin-based paints are not liable to attack on storage, but oleoresinous paints, once applied, provide food for ants and termites, and may be attacked by fungi to give acetic and oxalic acids and a range of higher monobasic and dibasic acids. Attack is concentrated at the resin-rich interfaces between coats so that symptoms include adhesion failures and flaking as well as discolouration, spotting and obvious mould growth. Paints on phosphated steel are especially prone to attack since phosphates are essential to the growth of most organisms. Microbiocidal additives used in paint formulations include barium metaborate, zinc oxide and cuprous oxide.

PETROLEUM PRODUCTS AND GREASES

In warm and humid atmospheres oil- and grease-based temporary protectives may sustain mould growth. Lanolin, a natural grease, derived from sheep's wool, is particularly prone to microbiological growth. Microbiological attack of soap-thickened greases results in attack of the soap with resulting loss of 'body' and rapid thinning of the film.

Bacterial growth is sometimes encountered in oil/water machining emulsions when they are used in recycling systems. Attack results in foul odours, breakdown of the emulsion, slime formation and rapid staining of metals after machining. Investigations have usually attributed these failures to the introduction of extraneous organic contaminants such as food residues and the products of animal pests, which provide part of the necessary nutrients. Hill (1968), in a survey of oil/water emulsions in use, identified biodegradation as a major cause of inefficient machining and subsequent corrosion. He recommended: (a) improved practice so that conditions do not encourage growth, (b) improved, biologically hard oil formulations, and (c) use of carefully selected biocides.

The mould, *Cladasporium resinae*, has given serious problems in jet engined aircraft. The mould grows in the aqueous phase, which inevitably forms in storage containers and in aircraft fuel tanks. Nutrients are derived from the fuel, and at the water/fuel interface fungal hyphae grow, and get carried into fuel lines and filters where they block the flow of fuel to the engines. Within the fuel tank where the fungus is in contact with the aluminium shell corrosion occurs, probably from the combined effects of passivation breakdown from oxygen removal, the physical shielding by the growing mass and the organic acid degradation products. The fungus is usually only encountered with aircraft and storage tanks that are used intermittently and in warmer climates. It is not a problem with commercial aircraft where a high rate of fuel throughput is maintained in both storage containers and aircraft fuel tanks. The risk is reduced by maintaining clean fuel stock, regular cleaning of tanks, good filtration, and arrangements for water separation and drainage from aircraft fuel tanks. 2-methoxy ethanol (2-ME) and organoborides have been recommended as biocides where the risk remains and their use has been evaluated by Elphick & Hunter (1968). However, Williams & Lugg (1980) report that although a concentration of 10% 2-ME in the aqueous phase is biocidal to *Cladasporium resinae*, bacteria (tentatively attributed to the genus *Pseudomonas*) have been isolated from both the aqueous phase at this concentration of 2-ME and the fuel in equilibrium with it. The bacteria produced a mixture of organic acids and caused the fuel and aqueous phases to emulsify. The carboxylic acids identified were succinic, a keto glutaric, citric and isocitric acids; these authors suggest that these acids will corrode aluminium. The bacteria survived in 25% 2-ME for eight days but did not appear to thrive at this higher level.

PLASTICS AND RUBBERS

Rubbers, and plastics based on natural materials such as cellulose or protein, are readily attacked by micro-organisms. Vulcanised natural rubber is also liable to attack by *Thiobacillus thio-oxidans* which derives its energy from oxidation of sulphur. Synthetic resins are generally much less likely to be attacked but many of the apparently immune polymers on closer examination have been shown to be susceptible to some degree of attack. Thus for instance Bengson & Gillis (1968) have shown in experiments on thin films of polyethylene that in contact with a mixed microbial culture there was sufficient attack under some conditions for pin-hole leaks to form. Comparatively inert resins will support algal growth when exposed to sunlight and wet conditions. Plasticisers in PVC are susceptible to attack resulting in loss of insulation, stiffening, discolouration and staining (a pink staining is effected by *Streptomyces rubrireticuli*). The effects are worsened by weathering (sun, heat and rain) which may lead to decomposition of the polymer chain; once started depolymerisation is autocatalytic and accelerates rapidly.

GLASS

Certain fungi grow on glass surfaces, drawing part of their nutrient from films of impurities and from the atmosphere but also etching into the glass. Because of this hazard, optical assemblies need to be stored in dry conditions in tropical environments.

REFERENCES

Bengson, M. H. & Gillis, J. R. (1968) Walters, A. H. & Elphick, J. J. (Eds) *Biodeterioration of Materials*. Elsevier Publ. Co., p. 99.

Elphick, J. J. & Hunter, S. K. P. (1968) Walters, A. H. & Elphick, J. J. (Eds) *Biodeterioration of Materials*. Elsevier Publ. Co., p. 364.

Hill, E. C. (1968) Walters, A. H. & Elphick, J. J. (Eds) *Biodeterioration of Materials*. Elsevier Publ. Co., p. 381.

Miller, J. D. A. (1980) *Br. Corros. J.* **15,** 2, 92.

Tiller, A. K. (1982) Parkins, R. N. (Ed.) *Corrosion Processes* Appl. Science Publishers, p. 115.

Williams, G. R. & Lugg, M. (1980) *Int. Biodet. Bull.* ISSN 0020-6164 16(4). Winter, p. 103.

10

Metal cleaning

Metal cleaning is a general term used to describe not only processes involving removal of oxide coatings, temporary protectives, machining oils, handling residues and other soil, but also etching, grit blasting and polishing processes where a degree of metal pretreatment or final conditioning is involved prior to the application of protective finishes. The selection of a cleaning treatment, or more frequently a series of treatments, depends on the type of soil present, the composition of the base metal and the process which is to follow. Fig. 10.1 summarises the scope of metal cleaning: the soils to be removed, the metals, the available cleaning treatments and the treatments which are to follows.

A detailed account of cleaning methods to remove all types of soil from all metals for the full range of finishing treatments is beyond the scope of this book and the reader is referred to the appropriate specifications and guides for detailed advice. DEF STAN 03/2 (1970; soon to be re-issued) gives details of cleaning methods and pretreatments for most metals. Cleaning and pretreatment of magnesium and its alloys are specified in DTD 911C (1964). BS 1133 contains extensive details on cleaning and drying metals prior to packaging. DG-8 gives general advice on cleaning and DEF 1234 specifies treatments for metals prior to packaging. The last two contain comprehensive listings of appropriate cleaning and pretreatment specifications. The *Transactions of the Institute of Metal Finishing* contain many authoritative papers on new and current methods of cleaning and pretreatments.

The present account is restricted to a discussion of the types of processes used, the underlying principles involved and a specific application, the cleaning of low-alloy steels before phosphating and painting, as an example of the approach, but appropriate variations are indicated for other base metals and other final finishes. Those aspects of cleaning which are important in avoiding early corrosion failures are briefly reviewed.

Properly cleaned metal surfaces, in the sense of metal surfaces having no

Range of commonly encountered soils
Salts Greases, oils Carbon/part decomposed organics, grinding, blasting and polishing residues
High molecular weight organic acids or salts and polymers (drawing, rolling or extruding lubricants)
Oxide (scale, rust, tarnish) and inorganic contaminants (S^-, SO_4^{2-}, Cl^-, NO_3^-)

Cleaning treatments
Organic vapour
Organic liquid/emulsifiers and two-phase cleaners
Alkaline solutions (with additives and/or cathodic polarization)
Abrasive blasting (dry or aqueous shot, grit or organics)
Tumbling, scouring, grinding, polishing
Hand abrasion (wire brush, abrasives, hammering)
Machining
Fused salts (sodium hydride)
Acid pickling (inhibited/anodic/cathodic)
Detergents

Metals
Steels, irons and cast iron
Corrosion-resisting steels ('stainless')
Aluminium and aluminium alloys
Copper and copper alloys
Nickel alloys
Titanium and titanium alloys
Magnesium alloys
Zinc and cadmium
Tungsten alloys
Molybdenum alloys
Silver

Subsequent treatments
Nil (most metals)
Temporary protection
Electroplating
Electroless nickel plating (steel and copper)
Galvanizing (zinc on steel)
Sherardizing (zinc on steel)
Calorizing (aluminium on steel)
Solder coating (tin/lead on copper/steel)
Metal spraying (many metals on steel)
Ion plating (many metals)
Sputtering/vapour deposition (many)
Phosphating (steel)
Oxidizing (steel, copper, magnesium, aluminium)
Anodizing (aluminium, titanium)
Chromate passivation (zinc, cadmium, copper, magnesium, stainless steel)
Benzotriazole (copper)
Smoothing (steel, copper, aluminium)
Adhesive bonding (most)
Paint and other organic coatings (most)

Fig 10.1 — Cleaning requirements, treatments, metals and subsequent treatments.

other atoms or molecules chemically or physically attached, are only prepared under high vacuum and they oxidise on contact with air. Although in the context of metal cleaning, the meaning of 'a clean surface' is generally understood, not surprisingly an accepted definition is lacking, but a practical definition is that:

> 'A clean surface is free of visible oxide or debris and of any contamination which may be a cause of staining or corrosion in subsequent storage, and it is suited to the process which is to follow'.

The last part of this definition allows the essential variation to meet often conflicting requirements such as the chemically smoothed aluminium surfaces needed for decorative anodizing; the physical burnishing to give a low friction sliding surface or a pleasing appearance; the newly treated, reactive and lightly etched surface needed for electrodeposition; the rough surfaces required to give adhesion for metal spraying and the somewhat less critical preparation needed for phosphate treatments which themselves etch the surface, and for which even a degree of selective recontamination from a wipe with a turpentine-wetted rag has sometimes been recommended for optimum coating formation.

Metal finishing specifications and treatment schedules acknowledge the interaction between cleaning methods and the processing which is to follow, by proposing appropriate cleaning treatments, at the least in a negative sense by specific exclusions. Cleaning specifications cater for the same interdependence by including variations to suit different finishing treatments which may follow cleaning.

Cleaning methods are selected on their ability to remove the expected types of contamination and to give a surface condition appropriate to the treatment which is to follow. Within the range of technical possibilities, costs and convenience will probably determine the choice. Dominating factors include the outage (downtime) costs, cost of chemicals, plant depreciation, space occupied, effluent treatment, heating, inspection, amount and quality of wash water, power and labour. Account must also be taken of the less readily quantified effects of process time, process complexity, value and size of the items being cleaned, flexibility of the process, extent of attack on the base metal, extent of corrosive spray and vapour, toxicity, flammability and odour.

TYPES OF SOIL

Cleaning is a key procedure in the protection of metals. Errors in cleaning may not be evident until processing difficulties are encountered in following treatments, or even worse, as failures in final processing, storage or service, when the costs of correction may be very high, the quantities involved will be considerable and the loss of goodwill may be of even greater importance. Cleaning is therefore an area where control and inspection are needed and short cuts are to be avoided. The types of soil to be removed may be considered in three categories: water soluble, inorganic and insoluble, and organic.

Water soluble

These are the most easily removed soils but they are also the most corrosive, and they should therefore be removed quickly. They may come from previous processing, atmospheric contaminants, heat treatment salts or handling residues.

Inorganic and insoluble

These include oxide scale, carbon, foreign metals, pickling smut, abrasion and polishing debris. Some of these materials may become embedded in the surface and are difficult to remove. Acid pickling, mechanical methods, agitation, gas scrubbing and etching are used to remove them.

Organic materials

These include hydrocarbons (oils, petrolatum) long-chain acids and salts (lanolin, greases and soaps), esters (fats, alkyd paints), and miscellaneous but less reactive materials which include waxes, long-chain alcohols, synthetic polymers and inhibitors. The organic salts are soluble in dilute alkali and the organic acids will be converted to soluble salts by alkalis. Treatment with hot alkalis will saponify esters to give organic salts and alcohols. The mixtures of salts, alcohols and hydrocarbons readily form emulsions in alkaline solutions; the waxes, synthetic inhibitors and polymers are more difficult to emulsify although heat and agitation will help. Mechanical abrasives may be used as an alternative method of cleaning or as an aid in alkaline cleaning. Many organic substances may be removed by solvents but salts of organic acids and polymers are best removed with alkalis, emulsion cleaners or mechanical methods.

BASE METAL

The chemical composition, fabrication history and metallurgical condition of the metal being treated are important in selecting cleaning treatments because they determine whether a metal will or will not react, how readily its oxide will dissolve and how it responds to later processing. Cleaning treatments are needed for a range of purposes in most of which the relative reactivity of the metal, its oxide, and soil will be critical. The aim of cleaning may be to remove organic contaminants without affecting the metal or its oxide coating, to remove the oxide scale without attacking the metal, or to remove some of the base metal. If metal is to be removed chemically, then the alloy composition is the dominant factor, but if mechanical polishing, machining, linishing, shot blasting or other mechanical treatments are to be used then hardness, which is dependent on both composition and metallurgical treatment, is important.

Other considerations in selecting treatment solutions are the need to avoid harmful side reactions such as pitting, selective attack at crevices and joints, stress corrosion cracking, and hydrogen embrittlement. This last effect is particularly insidious since there is no visible evidence until failure occurs under loads which may be as low as 30% of the tensile strength.

Stress corrosion and hydrogen embrittlement are fortunately confined to a limited series of alloys and specific environments, albeit alloys which have other very favourable properties such as high strength and resistance to general corrosion. The problems have been discussed in Chapter 5 but it is appropriate to consider them further here in the context of metal cleaning since correct processing at this stage can often avoid the risk of later failure; conversely ill-advised treatments can induce failure during treatment or leave the material in a susceptible condition.

High strength alloys or metals known to be susceptible to stress corrosion or hydrogen embrittlement should not be stressed during processing; an exception has to be made for close-coiled springs where it may be necessary to extend the spring slightly to allow access of treatment solution to all surfaces. Any localised surface tensile stresses which may have developed during quenching, forming or machining should be relieved by a suitable low temperature treatment. The temperature of stress relief has to be below that at which damaging metallurgical changes can occur. A suitable temperature can usually be selected from within the range 150–400°C; the corresponding treatment times will range from 1 to 10 hours. Even 150°C may be too high for certain aluminium alloys and case-hardened steels, however, and stress relief at lower temperatures may be necessary. Alloys with such limitations should be avoided if possible and if they are encountered care must be exercised in selecting manufacturing techniques which do not introduce tensile stresses. It is usually advantageous to introduce compressive surface stresses into metals known to be susceptible to either stress corrosion cracking or hydrogen embrittlement. These beneficial surface stresses may be induced by shot peening, abrasive blasting or surface rolling. Either of the first two processes may be introduced as a final stage in the cleaning and give the added advantage of avoiding a wet cleaning treatment. It should be remembered that if compressive stresses are induced into the surface there will be corresponding tensile stress below the surface layers which subsequent machining may reveal. After processing any deformation such as surface stamping of inspection marks and part numbers should be avoided since they introduce high residual stress, stress-concentrating notches and sites where contaminants may collect.

In processing alloys susceptible to stress corrosion, chemicals known to induce cracking should be avoided. Thus chloride-containing solutions should not be used in processing susceptible stainless steels, and copper alloys should not be treated in ammoniacal solutions.

RINSING

Rinsing is an important operation in aqueous processing both as a fundamental cleaning treatment and as a final and essential stage in all other aqueous processes. Because of its importance and its frequent neglect it is appropriate to consider it first.

Rinsing may be carried out by hosing, spraying or dipping; sufficient contact time must be allowed for salts to dissolve and for solutions to diffuse away from the region adjacent to the components. Processing efficiency is increased by heat and turbulence at the metal surface. The presence of crevices, blind holes and cavities in components being treated delays diffusion away from the surface and care is needed in selecting the process, water temperature, treatment times, drainage and drying procedures. With difficult components a final rinse in very dilute chromic acid may be advisable.

Immersion in successive tanks is the most readily controlled rinsing process. After immersion in a first rinse tank components should be passed through a second rinse and preferably a third. The build-up of contaminants must be limited either by a counter-current flow of fresh water through the series of tanks, or by periodically refilling the first tank and bringing it into re-use as the final tank. Organic contamination will not be removed by washing, and vigilance must be exercised in ensuring that organic films are not present on items passing through the rinsing system. Visual observation of the 'water break' on emerging items if consistently maintained is usually a sufficient precaution. The water should initially cling to an emerging surface thinning by drainage but not breaking away unevenly, or even worse, leaving isolated dry or wet areas, sure pointers to oil films on the surface.

Considering a system of successive rinse tanks, taking a typical strength of processing solution as 20% and assuming that 'carry-over' contamination is not allowed to exceed 1% at each stage, then the concentration of contaminants carried into three successive rinse tanks should be no more than 0.2%, 20 ppm and 0.2 ppm. Since natural waters typically contain up to 3000 ppm solids with 12–300 ppm chloride and 30–1000 ppm sulphate, the third tank, to be of any value, should contain deionised water. Adequate time must be allowed for diffusion to reduce the concentrations of solutions retained by roughened surfaces and within enclosures with restricted access. Diffusion rates may be increased by using hot water, which has the added advantage of hastening solution of solids, melting waxes and greases, decreasing the likelihood of comtaminants remaining absorbed (or chemisorbed), reducing surface tension and assisting drying. Agitation through stirrers, circulation jets or ultrasonic transducers is also helpful; the last is of particular benefit in removing solid debris and in assisting the transport of solution away from crevices and cavities. Rinsing is critical if prior processing has involved treatments with corrosive ions, and cross-contamination of other components in such solutions should be avoided by ensuring that items treated in halide solutions do not share their first rinse container with components treated in more benign processes. If water softening or deionising systems are used, care should be taken to ensure that they are functioning correctly. Deionising systems for instance, when nearing exhaustion, begin to release the less basic anions such as silicates, giving unexpected contamination at the final critical phase of the rinsing process. To avoid these effects the conductivity of the deionised water should be monitored and maintained below 4 mS/m. Although a final rinse with deionised water is not always necessary, it is generally worthwhile for the added ease of drying it allows and the absence of unwanted patterns such as the 'snail trails' that often affect subsequent phosphating and may show through a polished gloss finishing paint; before electrodeposition it is essential.

Hosing or spray rinsing may sometimes be preferred for its speed, simplicity and lesser risk of contamination build-up. The jets penetrate crevices and scour the surface and so often give efficient removal of retained residues, adsorbed films and debris. An array of jets is well suited to rinsing simply shaped items on moving production lines but again deionised water should be supplied for the final jets. Vigilance is required to ensure that jets do not get blocked and that water pressure is maintained. Hosing is best suited to large items of simple shape but the efficiency of the process is almost entirely dependent onthe skill and application of the operator. It is particularly appropriate, for large components such as steel plate or large diameter pipe which have been acid pickled, prior to dip rinsing, or as a final stage in the use of organic/emulsion cleaning.

Items from the final rinse which are to be processed further in aqueous systems should proceed to the next treatment without drying, while other items should be dried rapidly by circulation of clean warm air. Proprietary systems based on chlorinated solvents are available as aids to rapid drying but they are best avoided, especially for steel, aluminium or copper alloys since some decomposition of the halide may occur and inhibitors added as stabilisers are often involatile and give unwanted contamination of the final surface.

SOIL AND SCALE REMOVAL

Precleaning is the first stage and is intended to remove the bulk of the heavy soil. Most temporary protectives and residues of lubricating greases and machining oils are conveniently removed by vapour or solvent degreasing. Less easily soluble organic contamination such as drawing lubricants is removed by soaking in emulsifiable solvent cleaners or hot alkali. The latter will also remove paint residues and the more obstinate temporary protectives such as aged lanolin. Removal of adherent soil may be assisted by spraying with power washes. Occasionally remaining insoluble soil such as carbon smut may require abrasive treatment. Any rust may be removed at this stage by pickling either in cold hydrochloric acid or hot phosphoric acid. Phosphoric acid is to be preferred where possible as the residues are less aggressive and it should certainly be used on components with crevices or deep recesses.

ALKALINE CLEANING

When the bulk of the soil has been removed, final cleaning is accomplished by a brief dip in a strong alkaline cleaner, preferably with applied anodic current. The temptation to use the precleaning stage to achieve this final cleaning also should be resisted since the heavy levels of contamination which build up in the precleaning process make control difficult and are generally incompatible with complete soil removal. Immediately following the alkaline treatment items should be thoroughly rinsed.

ACID DIPPING

Immediately before phosphating any alkaline residues, oxide film and debris should be removed in dilute phosphoric acid and the items given a cold rinse and transferred to the phosphating bath without being allowed to dry.

DRY METHODS OF CLEANING

Light grit blasting is often preferred as an alternative to alkaline cleaning and acid dipping since it is faster and avoids time- and space-consuming solution processes. Phosphating solutions react rapidly with the grit-blasted surface to give a finer and more even coating than is obtained after aqueous cleaning. Grit blasting and other dry cleaning processes are particularly useful for processing steels susceptible to hydrogen embrittlement since they avoid the risk of hydrogen pick-up and the consequent time-consuming baking treatments. Final cleaning by grit blasting is also suitable before direct painting (preferably using a pretreatment primer as an initial coat) and is essential before most types of metal spraying but is of course not acceptable before electroplating, where tolerances are critical, or a smooth surface finish is needed.

OTHER METALS

Aluminium and zinc oxides are dissolved by alkaline solutions at pH values above 8.3 and 10.5 respectively. If dimensions are not critical then the consequent attack may be tolerated for the extra effectiveness of the cleaning that results, but for most purposes solutions based on sodium carbonate or silicate should be used with these metals to avoid attack.

Halogenated solvents in hot solvent degreasing baths (trichloroethylene, perchloroethylene or trichloroethane) tend to decompose to hydrochloric acid, especially in the presence of metals. Some metals, notably aluminium, zirconium, titanium and magnesium are especially active in initiating decomposition. Once decomposition begins the rate accelerates and quantities of hydrogen chloride, and other noxious and corrosive volatiles are evolved; the odour gives early warning of decomposition but the risk is slight if inhibited solvents are used and the supplier's recommended procedures are followed. These require that swarf is removed from all metals, especially the more reactive ones, before they enter the bath, and they stipulate that the solvent is changed regularly and the bath cleaned; periodic checks of the acid content of the solvent are usually advised. After degreasing all solvent should be removed from the metal since any that is trapped is liable to decompose and cause corrosion, especially on aluminium and brass. Hot trichloroethylene and hot perchloroethylene solvents may cause stress corrosion cracking of titanium/aluminium (5%)/tin (2%) and items made of this alloy should be heat treated to relieve any stresses before being exposed to these solvents.

Acid pickling is intended to remove contaminants and oxide from a metal surface without attacking the base metal; inhibitors are often added to reduce attack both to avoid loss of metal and to conserve acids. As an alternative to inhibitors, the item may be made cathodic so that attack is prevented and a degree of extra cleaning results from the 'scrubbing' action of the hydrogen bubbles evolved. Some treatments such as electroplating and anodizing require a higher standard of surface preparation, usually a light etch for plating and a chemical smoothing for anodizing. The ease with which metals are attacked by acids varies with the electrochemical potential of the metal and the protective character of the oxide film. In formulations for attacking metals which form strongly protective films such as aluminium, titanium and chromium, it is usually necessary to include a complexing agent which will form soluble complexes with the oxide; with less strongly passivating metals such as copper and nickel the presence of chlorides may be sufficient since they tend to render the oxide film porous and less protective. Minor phases rich in alloying additions are sometimes attacked at a different rate to the rest of the metal and this may give pitting, trenching or the disfiguring presence of undissolved residues, as smut on the surface, which is often difficult to remove. The physical properties of the solution may also affect the pattern of attack, so that viscous solutions based on concentrated phosphoric acid for instance are used to give smooth surfaces since convection currents do not replenish solutions within pits and crevices, and asperities are therefore preferentially attacked; the smoothing effects may often be enhanced by making the item anodic. Formulating acid solutions for pretreatments before the many finishing processes available for the wide range of alloys is a large and complex subject. Table 10.1 gives

Table 10.1 — Acid treatment solutions

Metal	Reagent	Temp. (°C)	Treatment
(a) *Removal of corrosion product with no attack on base metal*			
Steel	Sb_2O_3 20g $SnCl_2.2H_2O$ 50 g HCl (relative density 1.18) H_2O to 1 l	RT	Immersion with agitation until corrosion product removed
Copper and Brass	$SnCl_2.2H_2O$ 100 g HCl (relative density 1.18) to 200 ml H_2O to 1 l	RT	2 min, immersion with agitation until corrosion product removed
Aluminium and magnesium alloys	CrO_3 100 g H_2O to 1 l	90–95	2 min immersion with agitation until corrosion product removed
(b) *Pickling solutions (remove oxide with some attack on base metal: attack on metal may be prevented by cathodic polarisation)*			
Steel	(i) H_2SO_4 20% (ii) HCl 20% (iii) H_3PO_4 20%	10–60 RT 85–100	
Stainless steel	HNO_3 20% HF 15%	40–50	
Copper alloys	H_2SO_4 20%	30–50	
Aluminium alloys	CrO_3 1% H_3PO_4 1%	100	
	HF 4% HNO_3 14%	40	1 min immersion
Titanium alloys	HF 1.5–2.0% HNO_3 30%	20–65	
Nickel alloys	HF 2.0% HNO_3 30%	40–65	
Magnesium (fluoride anodizing: DTD 911C is preferred for magnesium)	CrO_3 10%	100	1.5 mins only
(c) *Etching and polishing solutions*			
Steel	$FeCl_3$ 20% HCl 10%	RT	(Used as etch before plating)
Stainless steel	HNO_3 30% HCl 8% H_3PO_4 12% Acetic acid 50%	70	Bright dip
Copper alloys	(i) H_2SO_4 60% HNO_3 20% H_2O 20%	RT	
	(ii) H_2O_2 1.5–2% H_2SO_4 0.2%	RT	Polishing rate 12 μm/h: final dip in 10% H_2SO_4 to remove the oxide film formed

RT, room temperature

some typical formulations to provide an indication of those used, but the reader is referred to the sources listed earlier in this chapter for detailed advice.

11

Conversion coatings

Conversion, or pretreatment, coatings, are applied to enhance the performance of subsequently applied organic protective coatings. They are sometimes used as protective coatings in their own right, particularly chromate treatments of aluminium, zinc, cadmium and copper alloys and anodized coatings on aluminium.

Conversion coatings are formed on the metal surface by reaction of the metal with the process solution, frequently by thickening the oxide coating, but sometimes also incorporating inhibitive anions such as chromates and phosphates into the surface film. The major pretreatments are: phosphating and black oxide coating of steel; anodizing and chromate conversion coatings on aluminium, magnesium and copper alloys; phosphating and chromate passivation of zinc, and chromate passivation of cadmium. Anodizing and chromate treatments are sometimes applied to stainless steels to improve their corrosion resistance by removing contaminants, smoothing rough edges and removing asperities, although no visible film is formed. Titanium may be anodized for the attractive colours which can be obtained by dyeing the anodic film, for anti-galling or to improve bonding of paint and adhesives, rather than for any increase in its already excellent corrosion resistance. Tannic acid-based conversion treatments for steel have occasionally found favour, but it is difficult to specify compositions and treatment conditions for these natural materials.

Pretreatment, etching or wash primers are combined phosphating and chromating treatments which incorporate a polymeric binder and an organic solvent; they are applied by dipping, spraying or brushing, and cure rapidly. A typical system is specified in DEF STAN 80–15. They are often used on bare steel, aluminium, zinc and copper as a pretreatment before painting. The properties of the coatings are more akin to those of conversion coatings than to priming paints and they are therefore included in this chapter.

This is perhaps an opportune point to reiterate the necessity for adequate preparation at all stages in metal protection. Conversion treatments will only succeed is the prior cleaning has been properly carried out. A major but unsung advantage of these processes is that they render any deficiencies in surface preparation very visible so that inspection is easier and quality control is improved.

STEEL

Phosphating

Phosphating, sometimes known as phosphatising or Parkerising, is the most widely used pretreatment for steel. It developed from phosphoric acid cleaning, when it was realised that the deposit that built up on steel processed in well-used baths had protective properties. The essential reactions are:

$$Fe\ (steel) + 2H_3PO_4 \rightarrow Fe(H_2PO_4)_2\ (soluble) + H_2\ (gas)$$

$$Fe + Fe(H_2PO_4)_2 \rightarrow 2FeHPO_4\ (insoluble) + H_2$$

Phosphating baths were developed using nearly saturated solutions of ferrous dihydrogen phosphate with a slight excess of phosphoric acid. This formulation functions at 90–95°C and items are immersed until visible hydrogen evolution ceases, usually after 35–45 minutes. A range of modifications and improvements have been incorporated since the process was first developed. Now manganese or zinc dihydrogen phosphates are usually used for the initial bath make-up; accelerators, usually nitrates or perchlorates, are added to allow reduced treatment times and lower bath temperatures; and spray processes have been developed. The situation has been clarified in the specifications BS 3189 and DEF STAN 03-11, which separate coatings into classes and give details of processing and coating characteristics.

Phosphating processes are in three main classes:

- Class 1 consists of manganese or sometimes iron phosphate giving heavy coatings of at least 7.6 g/m^2. They are usually unaccelerated;
- Class 2 accelerated baths, usually based on zinc phosphate which give a lighter weight of coating, minimum 4.3 g/m^2;
- Class 3 processes which give coating weights of 1.1–4.3 g/m^2. These are usually formulated for spray application.

(BS 3189 includes two coatings of even lighter weight for sheet steel.)

Best results are obtained on surfaces which have been mechanically abraded; grit blasting is particularly good and results in a fine even coating and shortened processing times. Items which have been cleaned by alkaline processes or inhibited acids should be immersed briefly in hot 20% phosphoric acid before treatment to remove any adsorbed inhibitor that might interfere with subsequent processing. After phosphating a final rinse in very dilute chromic acid or a mixture of phosphoric and chromic acid is often used to improve further the performance of phosphate coatings.

The strength of phosphating baths is determined by titration of a 10 ml sample against 0.1 M sodium hydroxide to a phenolphthalein end point; the quantity in millilitres taken to neutralise the bath sample is referred to as the the 'pointage' and is typically about 30. The proportion of the free phosphoric acid is also an important parameter and this can be measured in the same way as the pointage using methyl red, but the end point may be difficult to distinguish and is best found using a pH meter when free acid and total pointage can be determined by titrating to pHs of 4 and 9 respectively.

Class 1 coatings give the highest level of protection and are mostly used in conjunction with temporary protectives, or with lubricating oils on items subject to heavy wear such as gear wheels, shafts, heavy cutting blades, wearing surfaces and gun mechanisms, or on tools and machinery, especially on threaded or close-fitting components. Unaccelerated class 1 coatings are also occasionally used as a pretreatment for painting on creviced components where there is a risk of treatment solutions being retained after rinsing, since residues from unaccelerated baths are not corrosive, unlike many of those from class 2 and most of those from class 3 processes, which give corrosive residues.

Class 2 coatings are widely used as pretreatments before painting on rigid items. Nitrites, nitrates and occasionally nitroguanidine are the most frequently used accelerators. Any traces of nitrite or nitrate which are retained in crevices or blind holes are liable to be a cause of subsequent corrosion; retained residues of nitroguanidine are not corrosive.

Class 3 and thinner coatings are widely used as a pretreatment for painting on sheet steel which may flex, and when speed of processing is important. Nitrite, chlorate or perchlorate accelerators are widely used in these processes and any residues may be very corrosive so that good washing is vital.

Phosphate coatings greatly improve the corrosion protection of oil, grease and paint coatings both through their own protective action and by ensuring a high standard of surface cleaning, improving adhesion and inhibiting the spread of rust from points of damage. Fig. 11.1 shows the improvement obtained by phosphating before painting.

The improvement is greatest with the thick coatings and least with the class 3 coatings but the latter do not readily shatter if the surface bends, as the thicker coatings do, and so are tolerant to the demands made of items such as car bodies and steel drums. Phosphated high strength steels require de-embrittling baking treatments as outlined in Chapter 5. Andrew & Donovan (1970) showed that hydrogen pick-up during phosphating varied with the type of steel from 0.11 to 0.44 parts per million and that hardened tool steel was seriously embrittled by the processing; the hydrogen could be removed and the embrittlement was removed by baking at 200°C for one hour (see DEF STAN 03-4). This baking treatment does however slightly degrade the corrosion protection afforded by the coating. The effect is shown in Fig. 11.2.

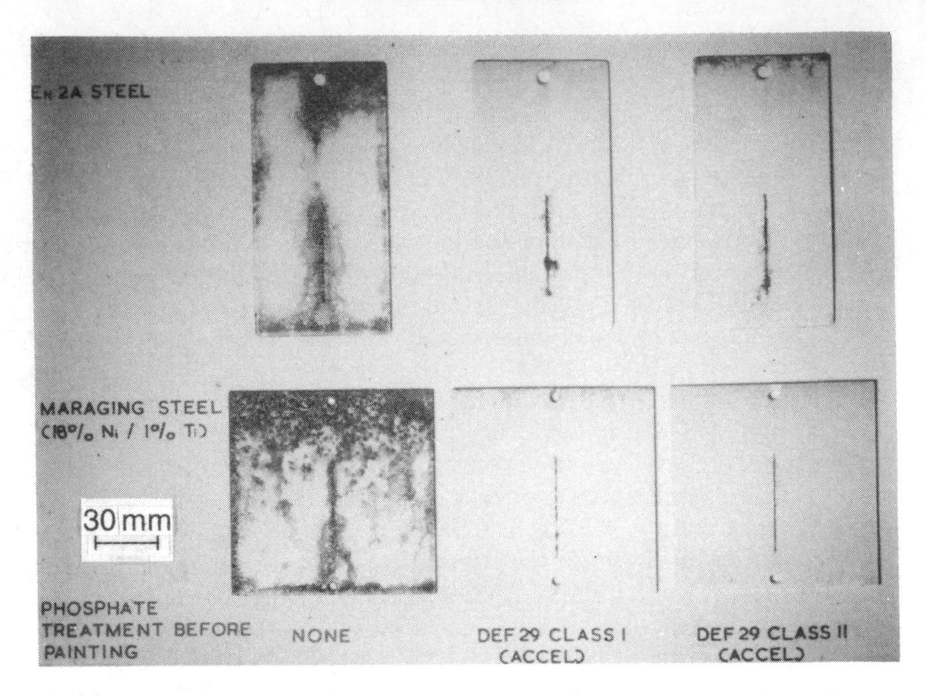

Fig. 11.1 — Painted steel panels after one month's salt droplet test (BS 1391): panels scratched and dented before testing. (© Controller, Her Majesty's Stationery Office 1986.)

Good phosphating requires the steel to be fully covered with crystals. A typical good quality coating from a class 1, unaccelerated manganese bath is shown in Fig. 11.3

Porous coatings may result through inappropriate cleaning treatments, high levels of free acid in the treatment bath or from segregation of alloying elements in the steel. The quality of the coating depends largely on the initial rate and pattern of attack. Zantout & Gabe (1983), for instance, in experiments on an unaccelerated zinc bath, showed that initial stimulation by slight anodic polarisation using a small current (0.5–2.0 mA cm^{-2}) for short times (0.5–2.5 minutes) gave high coating weights and low porosity; acceleration appeared to be obtained without the disadvantages associated with chemical accelerators.

Mechanism: experiments by Andrew and Donovan (1971) indicated that precipitation of the phosphate crystals occurred between the anodic and cathodic areas where the concentration of ferrous ions was increased by the anodic reaction and the pH was increased through the cathodic reaction:

Anode: $Fe \rightarrow Fe^{2+} + 2e$ i.e. increase in ferrous ion concentration
Cathode: unaccelerated $2H^+ + 2e \rightarrow H_2$ i.e. H^+ lost: pH increases
accelerated $O + H_2O + 2e \rightarrow 2OH^-$: pH increases.

Efficient deposition does not occur at fixed anodic or cathodic sites, and if the steel contains active areas such as manganese sulphide-rich inclusions

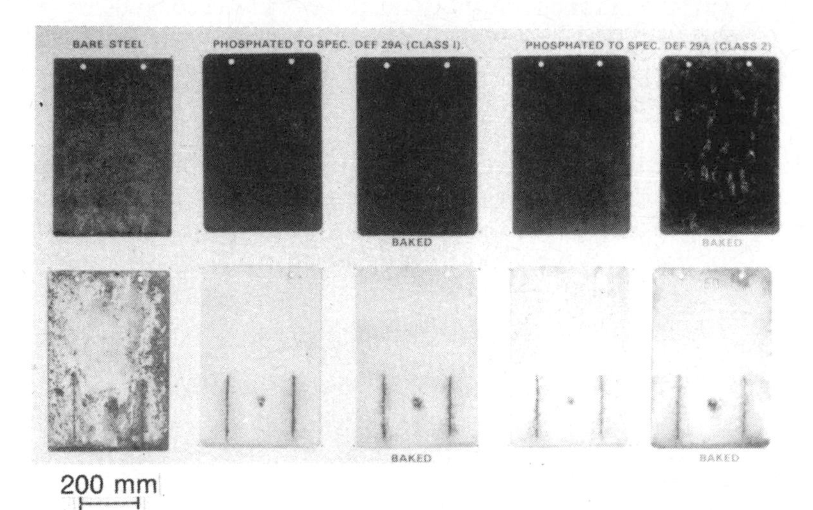

Fig. 11.2 — Effect of baking at 200°C for one hour on the performance of phosphate coatings, alone and with organic finishes; upper coated with lanolin, lower single coat of alkyd paint; 1 month salt droplet test BS1391 (1952). (© Controller, Her Majesty's Stationery Office 1986.)

Fig. 11.3 — Class 1 manganese/iron phosphate coating on steel. (© Controller, Her Majesty's Stationery Office 1986.)

which are efficient cathodes, so that a fixed pattern of cathodes and surrounding anodes forms, then there is a marked tendency to pitting. High free acidity increases the chance of pitting, while processes such as grit blasting or abrasive cleaning, which give a more even distribution of the electrochemical reactions and increase the rate of attack, reduce the risk of pitting.

The process of nucleation of the phosphate crystals and the factors that determine their initial adhesion to the substrate are difficult to investigate, but the available evidence suggests that it is probably a critical stage in the processing. Some workers have claimed that the crystal growth is epitaxial to certain faces of the grains of the substrate.

The nature of the accelerators and grain-refining additives is a critical factor which James & Freeman (1971) reviewed, showing how coating weight and the structures of the deposits can be varied over a wide range by additions of combinations of nitrates, nitrites, chlorates, perborates and peroxides to zinc phosphate solutions.

Oxide coatings

Thick oxide coatings can be developed on steel by treatment in baths of molten caustic soda with nitrate additions or by older processes involving heating in air followed by immersion in oil, sometimes through several cycles. Attractive blue, black or brown burnished coatings result. As a general pretreatment the process has been almost entirely replaced by phosphating and its use is mainly confined to the working parts of hand guns and shot guns on which, with a coating of light oil, it combines an attractive and traditional appearance with a degree of protection.

Although the caustic soda processing baths are free of water they give rise to hydrogen evolution:

$$Fe + 2NaOH \rightarrow Na_2FeO_2 + H_2$$

High strength steels therefore require baking de-embrittling treatments after processing. At least one fatal accident has resulted from a failure to appreciate this need.

Pretreatment primer (WP1)

One part of this two-pack system, also known as wash primer or etching primer, consists of polyvinylbutyral resin and zinc tetroxychromate in ethyl and *n*-butyl alcohol; the second part is phosphoric acid with ethyl alcohol. This formulation is covered by specification DEF STAN 80-15; a slightly different composition, 'T' wash, contains phenolic resin and is covered by British Rail Board Specification No. 62—Item 40. Pretreatment primer was developed for use on zinc, but experience has shown that it also performs well on steel and on aluminium. The phosphoric acid reacts with the metal providing a degree of *in situ* phosphating, and the chromate is an efficient inhibitor. The treatment improves the adhesion of paint and prevents the spread of rust from damaged areas almost as effectively as phosphating (see

Fig. 11.3). Alternative phosphoric acid-based single-pack formulations are also available but they do not appear to match the performance of WP1. Some hydrogen pick-up occurs during the reactions of etch priming paints on steel so that any high strength steels which have been treated will require baking to remove absorbed hydrogen.

ALUMINIUM ALLOYS

Aluminium alloys, especially those used in critical applications such as airframes and rocket motors, require a high standard of protection and this is often supplied by high quality synthetic resin-based paint schemes applied over a suitable pretreatment. These pretreatments, or modified versions of them, are also often used without paint, although sometimes with wax, in less aggressive conditions or sometimes on unalloyed aluminium on full exposure. It must be emphasised that aluminium alloys containing more than about 3% of copper or zinc are much more susceptible to corrosion than is pure aluminium or other aluminium alloys. The need of aluminium alloys for protection is especially dependent on the alloy composition and metallurgical condition. The major pretreatments are: anodizing, chromate conversion processes and etch priming. Etch priming is identical to the system used on steel and described in the previous section.

Thick oxide coatings can be developed on aluminium by anodic treatments in certain solutions. The process, known as anodizing, is used with varying treatment conditions to give coatings ranging in thickness from 1 to 2 μm up to 40 μm.

All aluminium alloys can be anodized although those high in copper develop thinner coatings which are porous and less protective. The coating forms uniformly over the surface even within holes and deep recesses. The coatings are generally colourless with thc exception of hard anodizing, which varies from brown to almost black, and the sulphuric acid process on certain alloys, especially those high in silicon, may give grey/brown coatings. Anodized coatings have found a range of uses. The boric acid process gives thin, flexible, non-porous coatings which are good electrical insulators and are used to provide insulation for electrolytic capacitors made from wrapped aluminium foil. The dense hard anodizing to BS5599 (1978) is brown or black and is obtained by anodizing in sulphuric acid at about 0°C; it is relatively expensive to produce but is valued in engineering applications for its high hardness and wear resistance. Chromic acid anodizing is used widely on coatings to assist in crack and pore detection since the coloured electrolyte is trapped in crevices and after processing it oozes out to give visible brown stains marking the position of cracks, pores or crevices.

The sulphuric acid process accounts for the vast bulk of commercial anodizing. The coatings provide a sound basis for paints or may be used as a final coating which can be made especially decorative by dyeing. A full range of bright colours, greys, browns and blacks can be obtained, including the popular golden yellow which looks like burnished brass. The chromic acid coatings may also be dyed but the colours are less intense. The coatings are sealed in boiling water after dyeing and this ensures that the dye is retained and the effect is permanent. The sealing treatment also increases the corrosion resistance of the coating which can be improved still further if sodium chromate solution is used for sealing, but this additional treatment degrades the colour of dyed coatings.

Retained residues of the sulphuric acid electrolyte will produce discolouration or more serious attack during the subsequent life of components. Therefore the chromic acid process should be preferred for creviced items or coatings. The chromic acid process is always used in anodizing surfaces which may be in contact with explosives or propellants as any risk of these materials contacting sulphuric acid residues is unacceptable.

In anodizing, the metallic aluminium of the surface is converted to the more bulky oxide and although some of the metal goes into solution there is an overall increase of volume and dimensions; the dimensions increase by about 25-40% of the thickness of the coating. The change is greatest with hard anodizing which is often used on sliding surfaces where fit is important, so that tolerances require special attention. This increase in dimensions may also close the mouths of any small holes or crevices in components with a consequent risk of trapping electrolyte.

Mechanism: the anodic film first grows through a charge transfer ion diffusion process between the double layer of charged ions; each 50 V potential gives approximately 0.1 μm growth of oxide by this mechanism. If the oxide is completely insoluble in the electrolyte, growth of the film then ceases. The thin-porous coating formed in boric acid is a typical example of this barrier type of coating. The thicker coatings are formed under conditions in which there is a slight tendency for the oxide film to dissolve in the electrolyte. The film grows initially by diffusion across the double layer but a pattern develops in which dissolution occurs at sites distributed uniformly over the surface. The oxide grows around these active sites so that they develop as pits and from their base oxide ions diffuse towards the metal surface and cause the film to grow on the surrounding areas. The growth mechanism gives the typically porous coatings of honeycomb appearance at high magnification as each growth site tends to grow independently to form a hexagon-shaped interface with its neighbours. In boiling water the outer layers of the oxide coating are hydrated and swell, sealing off the pores.

Xu, Thompson & Wood (1985) have carried out a comprehensive study on the mechanism of anodic film growth on aluminium. They identify a critical current density below which porous films grow but at and above which barrier films develop. For both types of growth Al^{2-} and O^{2-}/OH^- ions are mobile; whether one or other film develops depends on the fate of the Al^{3+} ions.

NON-FERROUS METALS (CHROMATE PROCESSES)

Chromate passivation

Chromates inhibit the corrosion of most metals; the chromate ion oxidizes any exposed metal and thus repairs and reinforces the protection offered by the air-formed film. Chromate treatments which produce a thickened and protective surface film have been developed for aluminium, magnesium, zinc, cadmium and copper alloys. These films typically have the yellow tint of the chromate ion often with the iridescent colours from optical interference; thinner colourless films and much thicker brown films may also be produced by varying treatment conditions and solution concentrations.

Chromate treatment of zinc and cadmium alloys gives protective iridescent films. The treatment consists of a 5–10 second dip in a 20% solution of sodium dichromate acidified with 1% sulphuric acid. The processing is covered by DEF–130, and Clarke & Andrew (1945) reported a detailed investigation of the processing of electrodeposited zinc with acid chromate solutions. The same solution improves the corrosion resistance of copper and brass although no visible film is formed. Clarke & Andrew (1961) modified the acid chromate solution by the addition of chloride and obtained iridescent films on copper and brass. Although these films give good protection for limited periods, the chloride ions retained within the film sometimes cause an unsightly and non-uniform thickening on prolonged humid storage.

A range of chromate treatments is available for spray, dip and brush application to aluminium alloys and gives iridescent, yellow and brown coatings. They confer improved corrosion resistance and are sometimes used alone but more often they are used as a pretreatment before painting. The process solutions include alkaline chromates and acid chromates; the latter are often modified with additions of phosphoric and sulphuric acids, and fluorides and nitrates. These treatments are much less expensive than anodizing, relatively tolerant to imperfections in surface preparation and provide a good base for paints, and therefore find wide applications.

Three classes of chromate are widely used in the protection of magnesium alloys: the acid chromate bath, the chrome manganese bath and the hot alkaline chromate bath. All three processes are included in the Ministry of Defence specification DTD 911C, and the compositions of the treatment solutions given there are:

- the hot half-hour chromate bath, which is recommended for all types of magnesium alloys where a high degree of protection is needed with negligible dimensional change, consists of a solution in water of:

ammonium sulphate	3.0% w/v
ammonium dichromate	1.5% w/v
potassium dichromate	1.5% w/v
ammonia solution (specific gravity 0.880)	0.27–0.43% v/v

immerse for 30 minutes in boiling solution;

- the acid chromate bath, which is particularly suitable for magnesium/manganese alloys or generally for unmachined parts to be kept in store. It is not recommended for parts machined to fine tolerances. The bath consists of a solution in water of:

	sodium dichromate ($2H_2O$)	15% w/v
or	potassium dichromate (anhydrous)	15%w/v
	nitric acid (specific gravity 1.42)	20–25% v/v

immerse for 10–120 seconds at room temperature;

- the chrome–manganese bath for all types of magnesium alloys where a high degree of protection is required with negligible dimensional changes. The bath consists of a solution in water of:

sodium dichromate ($2H_2O$)	10% w/v
manganese sulphate ($5H_2O$)	5% w/v
magnesium sulphate (H_2O)	5% w/v

immerse for variable times from room temperature to boiling: 1.5 hours at 20–30°C, 3–10 minutes at boiling point.

Etch primer, which was described earlier in this chapter, is used as a pretreatment on aluminium, zinc, cadmium, copper and steel before painting. It is the only reliable pretreatment for zinc if oleo-resinous paint is to be used and prolonged exposure to humid conditions may be required.

Chromate treatments are also available for silver, tin, titanium and stainless steels.

REFERENCES

Andrew, J. F. & Donovan, P. D. (1970) *Trans. IMF* **48,** 154.
Andrew, J. F. & Donovan, P. D. (1971) *Trans IMF* **49,** 162.
British Rail Board Spec. No. 62—Item 40.
Clarke, S. G. & Andrew, J. F. (1945) *J. Electrodep. Tech. Soc.* **20,** 119.
Clarke, S. G. & Andrew, J. F. (1961) *Proc. First International Cong. on Metallic Corrosion.* Butterworths, London, p. 173.
James, D. & Freeman, D. B. (1971) *Trans. IMF* **49,** 79.
Xu, Y., Thompson, G. E. & Wood, G. C. (1985) *Trans IMF* **63,** 98.
Zantout, B. & Gabe, D. R. (1983) *Trans. IMF* **61,** 88.

12

Permanent coatings

TYPES OF COATINGS

The major classes of permanent coatings used in protecting metals are:

Organic coatings
- Paint
- Lacquer
- Varnish
- Plastic (hot dipping, fluidised bed)

Metal coatings
- Hot dip
- Metal spray
- Electrodeposited
- Cladding
- Hot diffusion
- Impact
- Vacuum deposited (chemical, physical and ion implantation)

Inorganic coatings
- Anodic conversion coatings
- Enamels
- Zinc silicate paint
- Flame sprayed

The major mechanism through which these coatings afford protection are: physical exclusion of the environment by the inorganic and metal coatings, with additional sacrifical protection by the anodic metals; and exclusion of solids and ionic impurities by organic coatings, sometimes supplemented by inhibitors (priming paints), or by sacrificial protection (zinc-rich paints).

Anodic metals give the most complete protection while the coating lasts but general corrosion starts when the sacrificial metal is exhausted; the duration of protection increases with the thickness of the coating. Aluminium, zinc and cadmium on steel, zinc on aluminium, and zinc, cadmium or chromium/nickel on copper alloys are the most frequently encountered sacrificial coatings. Results of exposure tests by Clarke & Longhurst (1962) on both sacrificial (anodic) and unreactive (cathodic) metal coatings on steel are shown in Fig. 12.1. The extent of attack on these coatings is given in Table 12.1.

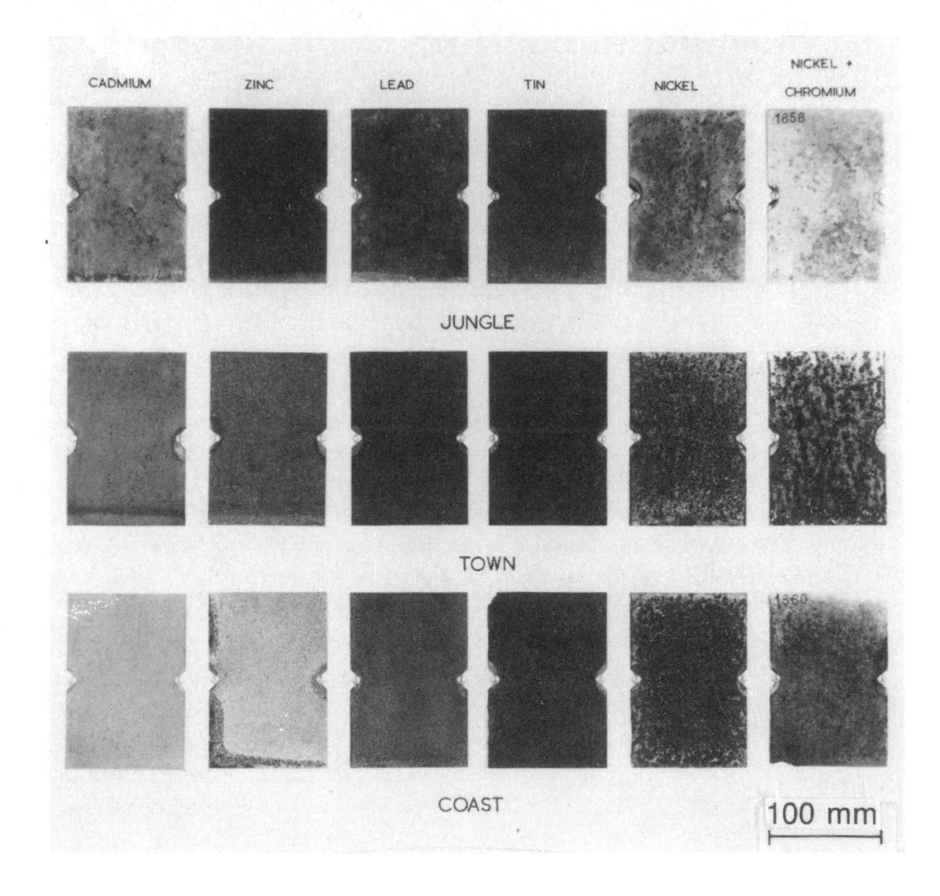

Fig. 12.1 — Metal coatings (12 μm) on steel after two years of exposure. (©Controller, Her Majesty's Stationery Office 1986.)

In the trials of Fig. 12.1 and Table 12.1, the anodic coatings zinc and cadmium provided good protection although the coatings suffered some corrosion; the cathodic nickel, and chromium on nickel, coatings provided poor protection although the coatings themselves suffered little attack. In contrast, the same two coatings gave good protection to brass in Fig. 12.2

Table 12.1 — Attack (μm) on 12 μm electrodeposited metal coatings on steel during two years of exposure in jungle, tropical town and tropical surf beach (test panels shown in Fig. 12.1)

	Cadmium (Cd)[c]	Zinc (Zn)[c]	Lead (Pb)[b]	Tin (Sn)[b]	Nickel (Ni)[b]	Chromium on nickel (Ni+Cr)[a,b]
Jungle	0.8	1.1	0.8	0.4	0.05	<0.05
Town	2.4	2.9	1.8	2.0	0.2	<0.2
Coast	0.7	5.7	0.7	6.0	0.45	<0.45

Notes: [a] 0.5 μm of chromium on 12 μm of nickel
[b] Steel rusted
[c] Steel unattacked while coating remains; rusting has started on steel protected by zinc on the coast where zinc has corroded through

(from the work of Clarke & Longhurst (1962)), since the base metal is less reactive, but also because the nickel is now anodic to the base metal.

Chromates, lead salts and oxides, and metal phosphates are the inhibitors most frequently encountered in priming paints and conversion coatings; benzotriazole is an effective inhibitor in lacquers for copper and brass. Inert undercoats and resin-rich topcoats provide physical protection and as shown by Mayne (1976) appear to act as molecular sieves, trapping reactive ions and maintaining a high resistance to ionic currents. Mayne (1976) also showed that polymer coatings are readily permeable to oxygen and water so that they do not reduce the rate of the cathodic reaction and if active contaminants are retained beneath the film corrosion occurs; the effect of such contaminants may be partly offset by inhibitors in the paint which act to polarise the anodic reaction of the corrosion couple.

The protective value of barrier coatings may be limited by the presence of minute pores (in paints known as 'holidays'), inclusions, cracks or other damage in the coating. The effectiveness of organic coatings is partly determined by the ease with which salts diffuse through the coating.

The choice of a finishing scheme is influenced by the need for corrosion resistance but is also determined in part by other requirements such as the need for resistance to wear, impact, heat, solvents and chemicals, but also and often most importantly, for its appearance. Defence Guide DG–8 gives detailed advice and cites specifications for most processes and materials used in protecting metals and is generally applicable to civil equipments, as well as defence, especially when protection in arduous environments is sought.

PROTECTION OF STEEL

BS 5493 (1977) gives valuable practical directions on all aspects of protecting steel. The Department of Industry Guide No. 12 (1982) contains useful summary advice on the selection and performance of paint for structural steelwork.

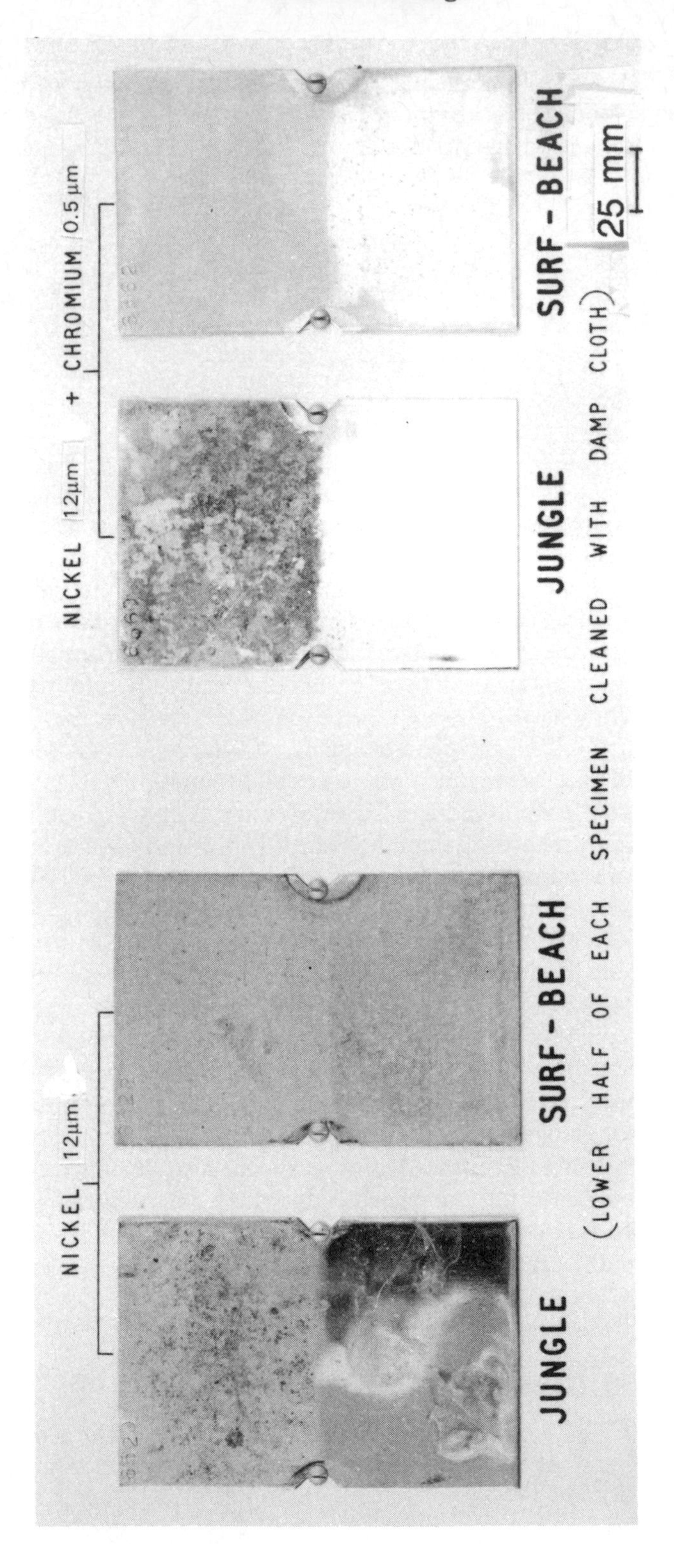

Fig. 12.2 — Ni and Ni/Cr coatings on brass after two years of exposure. (©Controller, Her Majesty's Stationery Office 1986.)

The sacrificial metals, aluminium, zinc and cadmium, protect steel effectively while the coating remains intact; their life to this point is determined by coating thickness; their performance is improved by painting. Cathodic coatings such as decorative nickel/chromium, or the hard and wear-resistant thick chromium, may give a lower level of protection which may be increased by wax or greases. These metals themselves are very resistant to corrosion but once attack begins on the base metal at pores, edges or cracks it usually extends laterally along the interface between the steel and coating, and since the coating presents a large area which may act as a cathode, corrosion may then be accelerated to higher rates than would occur on an unprotected surface. The performance of nickel and nickel/chromium has been much improved by the development of micro-cracked chromium (see Postins & Longland (1971)) and by duplex nickel: the mechanisms are illustrated in Fig. 12.3.

Paints are the most widely used coatings on steel and deservedly hold their popularity for their ease and versatility of application. Their performance depends critically on the cleaning and pretreatment of the metal before painting. The performance of paint on steel may be greatly enhanced by pretreatment of the steel by conversion coatings or metallic coatings applied prior to painting. The performance of paint on pretreated steel exposed for two years in a corrosive environment is shown in Fig. 12.4 (from Clarke and Longhurst (1962)).

A complete and even thickness of the paint coating is needed for efficient protection. This is relatively easy to achieve on fully exposed surfaces by spraying, dipping or brushing, and then drying in well-ventilated conditions, but difficulties are sometimes encountered. In enclosures for instance a cycle of solvent evaporation and condensation sometimes drains the paint or lacquer from the upper surfaces to leave them unprotected. A typical example of this phenomenon, known as solvent washing, is shown in Fig. 12.5. The lacquer coating on the phosphated steel component was thinned on the upper surfaces of the cavity by solvent washing during stoving. In subsequent storage the poorly protected area was seriously pitted by contact with the filling contained within the cavity, while the lower part remained uncorroded. Solvent washing affects paints, lacquers and solvent-based temporary protectives, and is one part of the general problem of protecting surfaces within cavities. This problem is compounded by the difficulties of inspection, so that mechanical failure or visible penetration to the outside are sometimes the first evidence of any deficiency (a chapter should perhaps be devoted to protection within cavities, but there are few satisfactory answers, except to remove difficult cavities at the drawing board).

Paints based on epoxide and polyurethane resin media give a higher standard of protection than the traditional drying-oil based paints. The synthetic resins possess the additional advantage that they do not cause corrosion of metals in enclosures (see Table 6.8) unlike the drying-oils which release volatile organic acids which may corrode nearby metals.

If cleaning is inefficient and rust remains, zinc-rich paint, or a primer containing red lead, zinc phosphate or chromates is to be preferred, but the

(a)

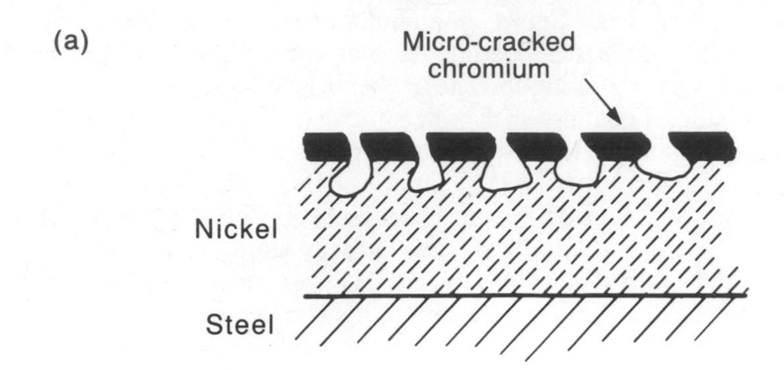

(b)

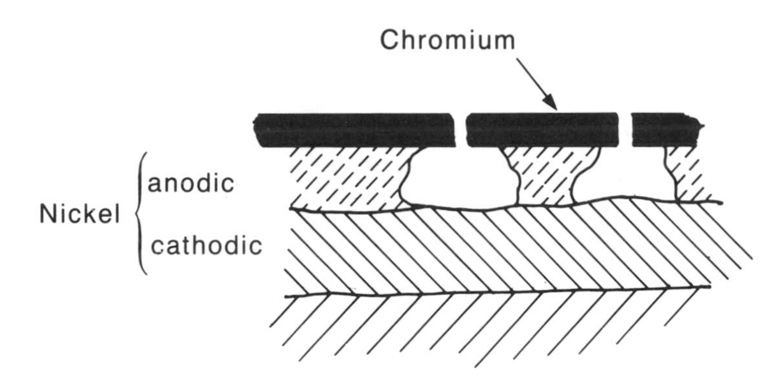

(c)

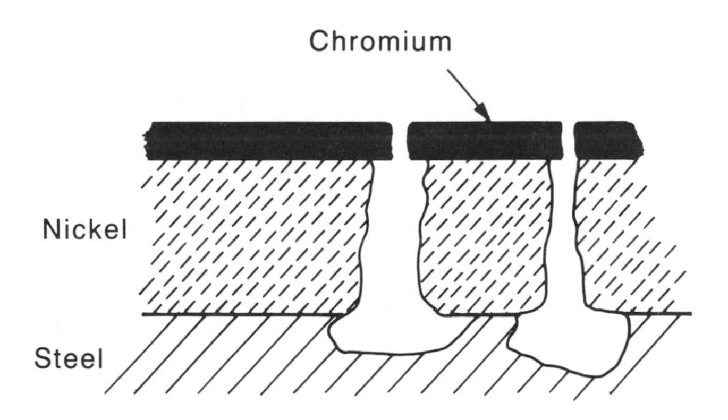

Fig. 12.3 — Reduction of corrosion penetration by (a) micro-cracked chromium and (b) duplex nickel. (c) Corrosion of steel under 'traditional' chromium on nickel. (©Controller, Her Majesty's Stationery Office 1986.)

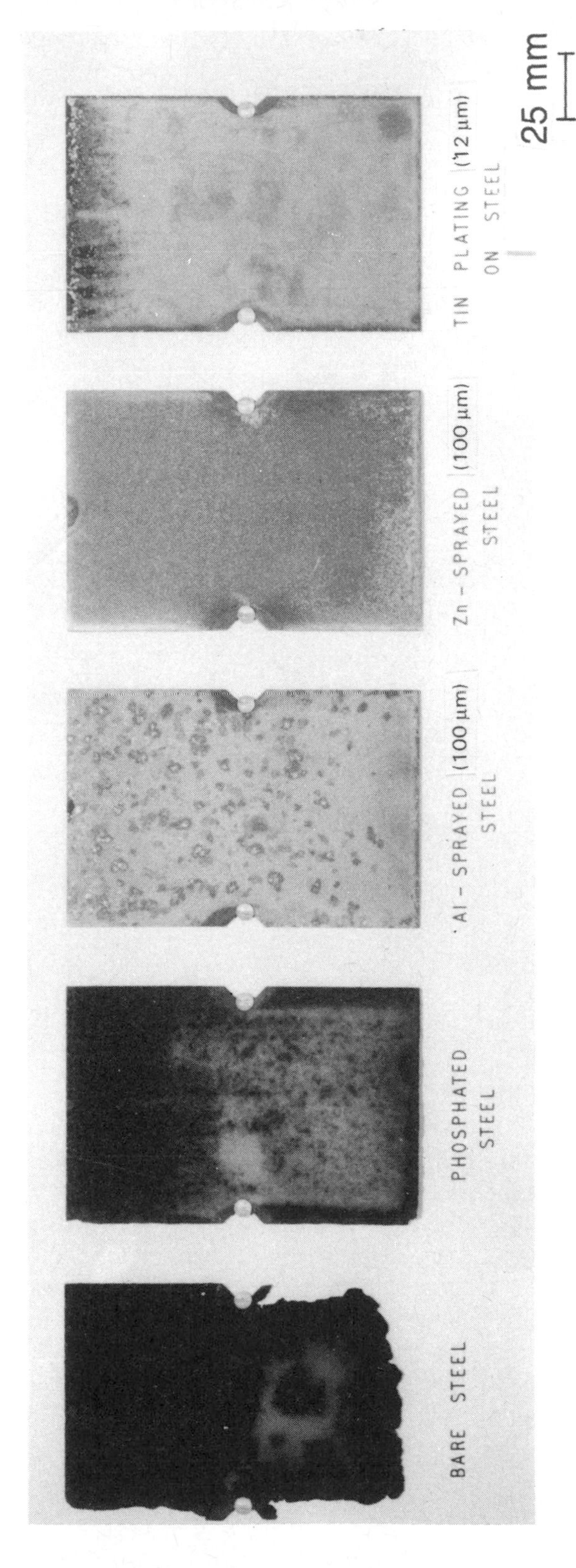

Fig. 12.4 — Paint on different substrates on steel after two years of exposure on a surf beach. (©Controller, Her Majesty's Stationery Office 1986.)

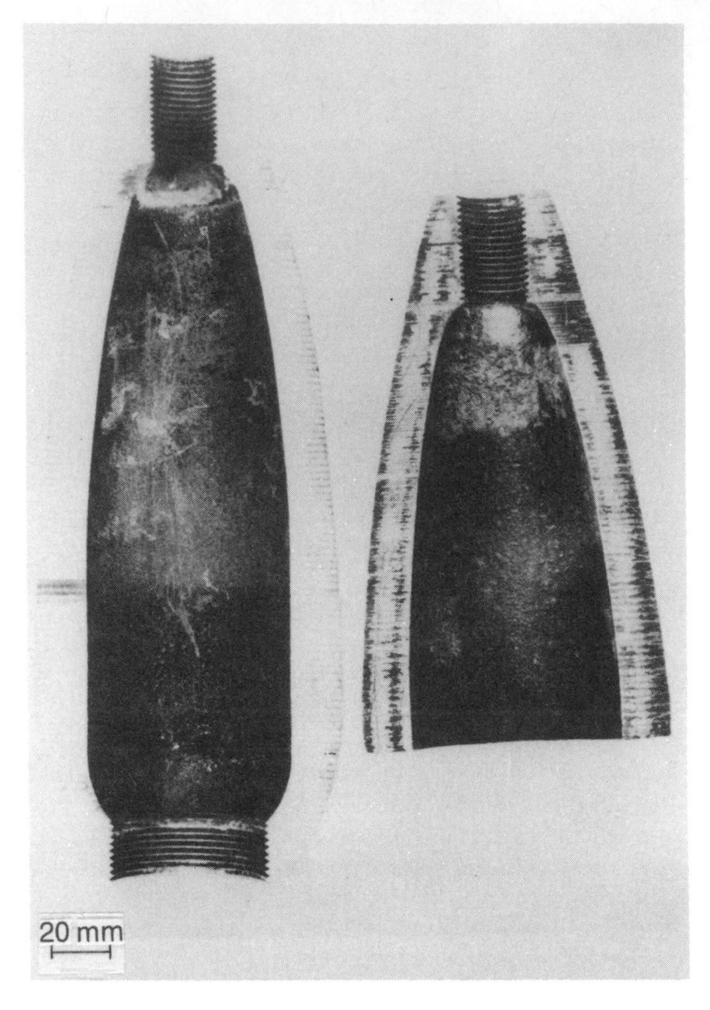

Fig. 12.5 — Thinned coating and lacquer failure due to solvent washing. (©Controller, Her Majesty's Stationery Office 1986.)

life of the coating will be reduced by the presence of rust, especially if it has developed in an aggressive environment and therefore contains a high proportion of sulphate, chloride or other active anions. Treatment with phosphoric acid helps to remove some of the soluble anions and also gives a degree of phosphate pretreatment; both effects improve the performance of the final coating. The Department of Industry Guide No. 13 (1982) gives a good account of the task of preparing all surfaces, including steel, for painting. A summary of the protection expected from coatings on steel is contained in Fig. 12.6 (from DG–8).

PROTECTION OF ALUMINIUM ALLOYS

Aluminium corrodes only very slowly in most rural and inland urban environments, developing a white/grey oxide patina, but is attacked more rapidly in marine areas; alloys, especially those of high copper or zinc content, corrode much more readily; some suffer from the very damaging layer corrosion and stress corrosion (see Chapter 5). The more susceptible alloys are sometimes protected by cladding with pure aluminium, the cladding being applied by hot rolling so that a secure bond is obtained. Alternatively a protective coating of zinc or aluminium may be flame-sprayed onto aluminium surfaces, and for best protection, is then painted. Anodized aluminium is widely used as a decorative cladding on the faces of modern buildings in inland cities and towns where, if a pure grade is used and it is properly anodized and sealed, and preferably cleaned regularly, it resists corrosion for the planned life of the building although care is needed to ensure that contaminated water does not drain over the surface.

The effect of two years of exposure on a surf beach on unprotected aluminium at two levels of purity, a cast aluminium alloy, an anodized aluminium alloy containing 4% copper and an alloy of magnesium is shown in Fig. 12.7 (from the work of Clarke & Longhurst (1962)).

A summary of the expected performance of coatings on aluminium is contained in Fig. 12.8 (from DG–8).

PROTECTION OF OTHER ALLOYS

The susceptibility of other metals and alloys on exposure outdoors together with the performance of various protective coatings is summarised in Fig. 12.9 (from DG–8).

REFERENCES

Clarke, S. G. & Longhurst, E. E. (1962) *Proc. 1st Int. Corros. Cong. (London)*. Butterworths. (The panels in Fig. 12.1, 12.2, 12.4 & 12.7 were part of this work although they are not shown in the paper.)

Mayne, J. E. O. (1976) Shreir, L. L. (Ed.) *Corrosion* **2,** 15, 30.

Postins, C. C. & Longland, J. E. (1971) *Trans. Inst. Met. Fin.* **49,** 89.

Corrosion Control No. 12, HMSO.

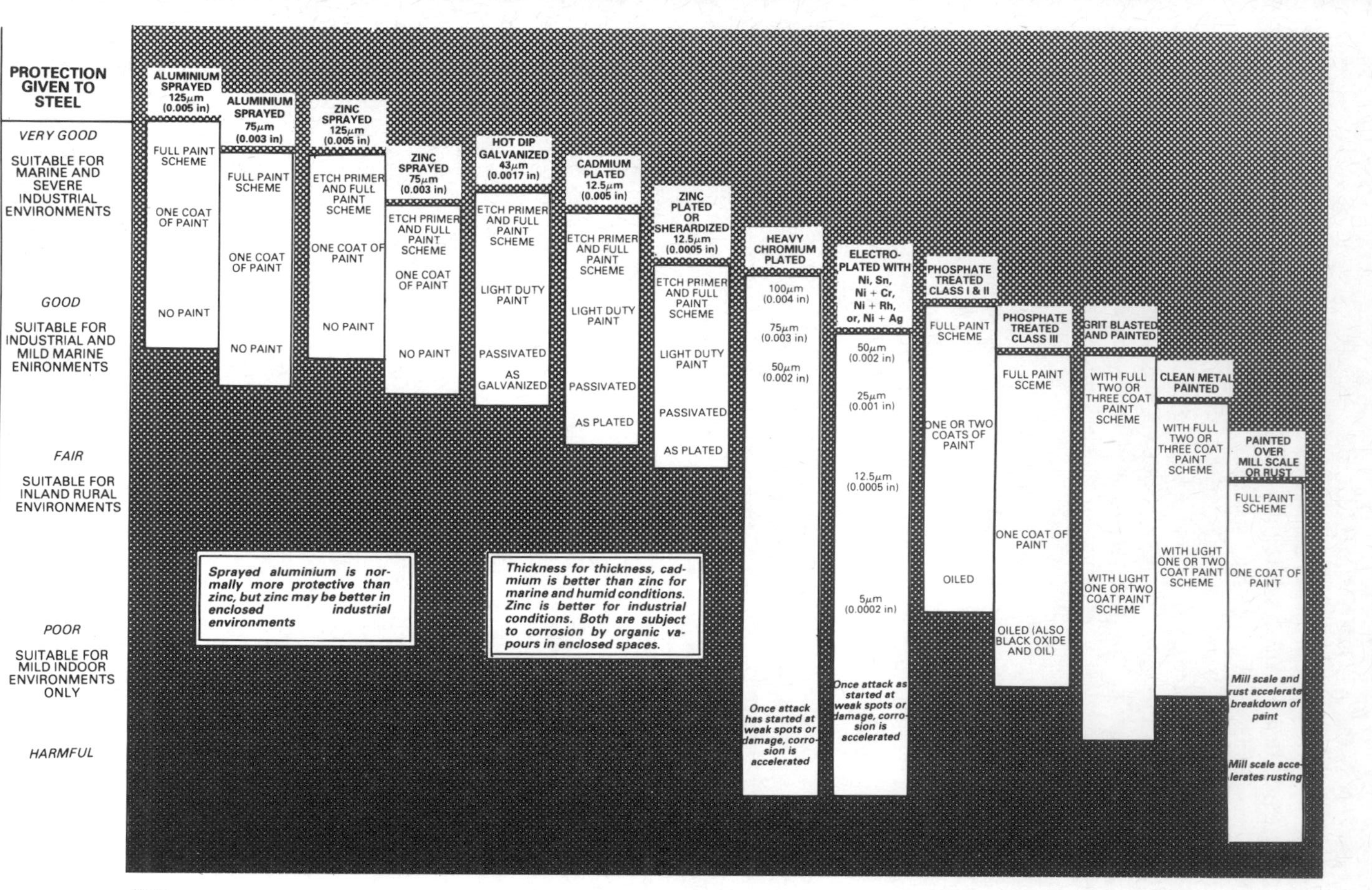

Notes

The chart relates to atmospheric exposore. It does not cover behaviour at elevated temperatures or special atmospheres e.g. vapour corrosion. The performance is related to the severity of the environment and is indicated by shading graded from light to dark.
Very Good: (No shading) Protective schemes required for marine and severe industrial exposure. They will give prolonged protection (10–20 years) under less severe conditions. *Good:* Protective schemes for most outdoor environments, but life limited (2–4 years) under severe conditions. *Fair:* Light duty protective schemes suitable only for mild inland rural and indoor environments. *Poor:* Treatments giving little or no protection, suitable for dry indoor environments. *Harmful:* (Darkest areas) Certain metallic coatings and mill scale accelerate rusting in corrosive environments once they have ceased to act as excluders.

Fig. 12.6 — Performance of protective coatings on steel. (©Controller, Her Majesty's Stationery Office 1986.)

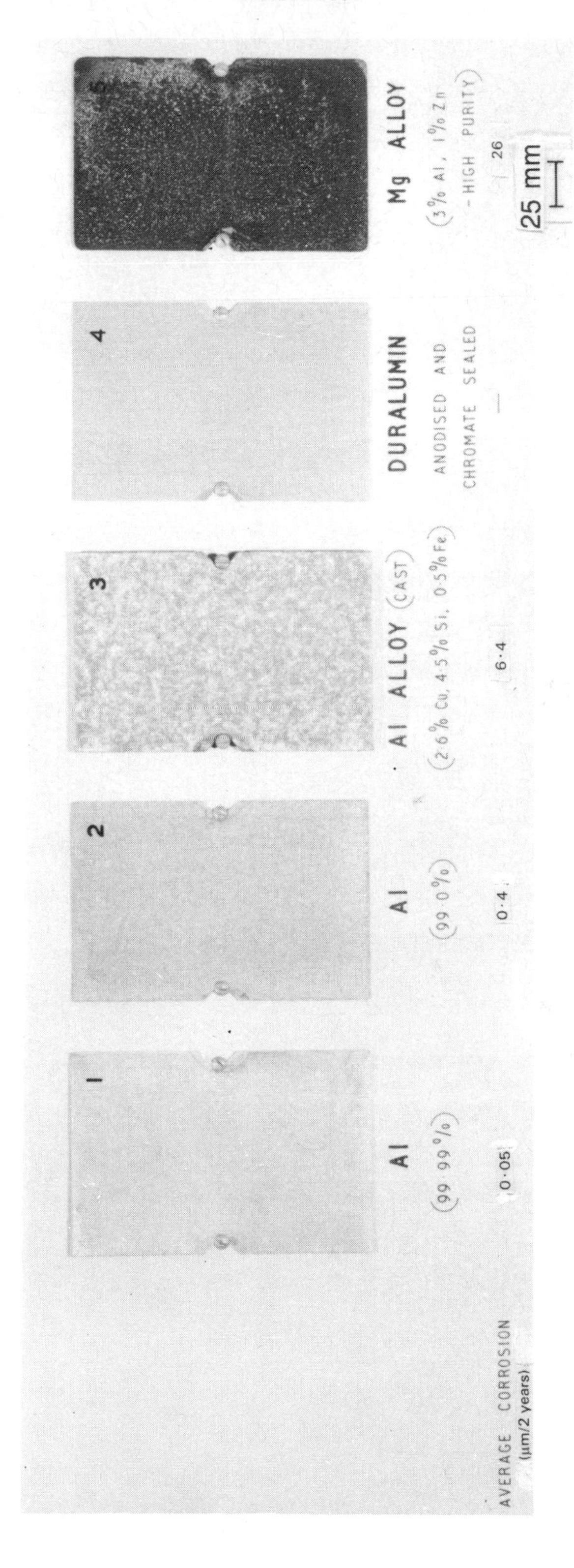

Fig. 12.7 — Light metals: two years of exposure on a surf beach. (©Controller, Her Majesty's Stationery Office 1986.)

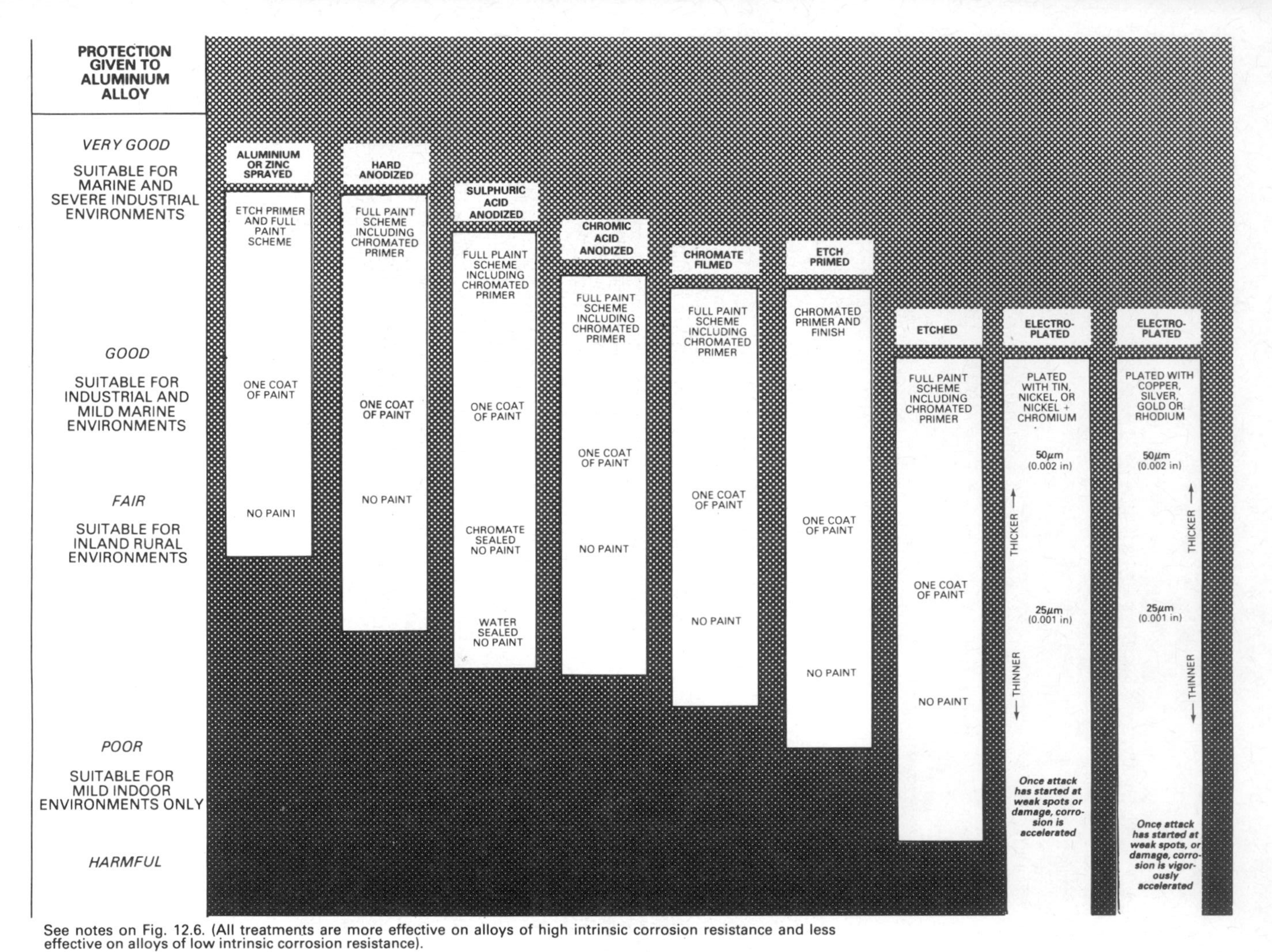

See notes on Fig. 12.6. (All treatments are more effective on alloys of high intrinsic corrosion resistance and less effective on alloys of low intrinsic corrosion resistance).

Fig. 12.8 — Performance of coatings on aluminium and aluminium alloys. (© Controller, Her Majesty's Stationery Office 1986).

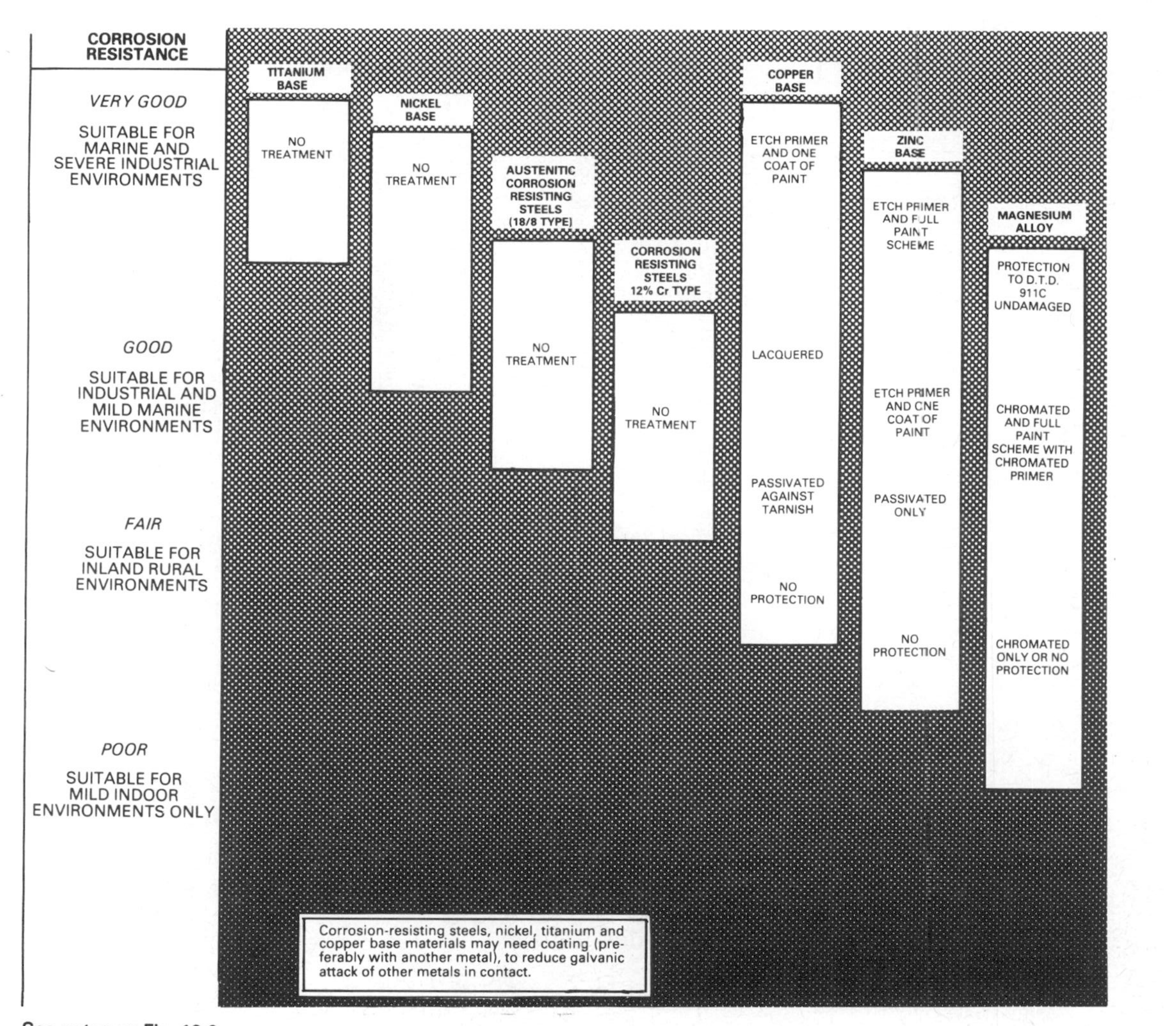

See notes on Fig. 12.6.

Fig. 12.9 — Performance of coatings on non-ferrous metals (excluding aluminium). (©Controller, Her Majesty's Stationery Office 1986.)

13

Temporary Protectives

Temporary protectives are 'temporary' in the sense that they can easily be removed without deleterious effects to the metal or to any permanent protective. The period for which they will continue to protect an item may vary from a few weeks to many years. They are used at all stages of manufacture from the production of the ingot, plate, sheet, forging or casting to the point of delivery to the final user, and indeed in some instances to the end of the useful life of the component.

Temporary protectives serve in three major roles: protection in inter-stage manufacture; to supplement the final finish between manufacture and arrival at the selling point, and as the sole protective on the surfaces of newly manufactured items which are to function in a protected environment in use. Examples of this last are the internal surfaces and components within engines and gearboxes, sliding and working surfaces of tools and machinery, and threads of nuts, bolts, pipes and pipe fittings. In these instances the protective is generally chosen to be compatible with the final working environment so that removal is not needed. It is important that temporary protectives should be easily removable when they are used to protect the final finish so that goods can be delivered to the customer or final user in 'new' condition although they may possibly have been stored for long periods in exposed conditions or may have been transported half way round the world. The most obvious example here is the motor car, but most large manufactured items have similar requirements.

British Standard 1133 (Section 6) gives a useful general summary of the properties and uses of the most widely used classes of temporary protectives; the BS classification is given in Table 13.1 with a brief summary of their uses and properties. The BS classification can include a wide range of compositions and properties, and products offered by individual manufacturers may be based on other criteria such as appearance or ease of removal, which the manufacturer considers has additional appeal or are designed for particular specialist requirements such as compatibility with engine oils or

Table 13.1 — BS 1133 Classification of temporary protectives

Type	Methods of application	Ingredients
TP1 Thin hard film — solvent deposited TP1a quick drying TP1b slow drying TP1c slow drying, and water displacing	Dip or spray	Solutions of protectives such as plasticized resins in volatile solvent
TP2 Thin soft film — solvent deposited TP2a ordinary grade TP2b water displacing	Dip or spray	Solutions of protectives such as lanolin in volatile solvent
TP3 Thick soft film — hot dipped	Dip	Usually based on petrolatum
TP4 Soft film grease Tp4a soap/oil grease TP4b castor oil/lead stearate	Brush or smear	4a: a metallic soap mineral oil-based grease 4b: compatible with rubber
TP5 Medium thickness soft film (semi-fluid)	Brush or swab	Inhibitors (e.g. wool fat derivatives) in mineral oil, sometimes with petrolatum
TP6 Oil, thin film	Dip, spray or circulate	Lubricating oils with soluble inhibitors
TP7 Thick strippable hot-dip coating	Dip	Ethyl cellulose with mineral oil, plasticisers, resins and stabilisers.
TP8 Medium strippable cold applied coatings	Spray, brush or dip	Solutions of protectives such as vinyl copolymer resins, plasticisers and stabilisers. Inflammable and non-inflammable solvents
TP9 Volatile corrosion inhibitor	On paper wrap, as powder, or in solution	
TP10 Contact inhibitor	On paper wrap	

special formulations intended to combat particular environments. Readers should beware of some of the more extravagant claims for materials which are recommended for use on already contaminated or corroded surfaces.

The Ministry of Defence guides and specifications give comprehensive information on the use of temporary protectives. DG-8 lists the temporary protectives in use in the Ministry of Defence and gives their field of use, application, removal, precautions and limitations. DEF-1234A and DG-11 (1976) give specific information on the use of temporary protectives in packaging. DG-12A gives details of all petroleum products in defence use.

The Institute of Petroleum coordinates the standardisation and classification of test procedures for petroleum products and their handbook gives those for temporary protectives and other petroleum products.

Beale (1976), in a review of the applications and uses of temporary protectives, has included a valuable bibliography of previous publications, and the Department of Industry's Guide No. 6 (1981) gives a good practical introduction to their properties and uses.

Generalisations on temporary protectives are of limited value since their only unifying characteristics are that they are temporary, protect metals and are mostly formulated from petroleum products. The approach in this chapter is based on user evaluations by the writer and his colleagues designed to quantify the protection that the products in defence specifications afford, and on experience in their use. The performance of temporary protectives depends on the conditions of use, which, in storage and transit with and without various standards of packaging and air conditioning, vary so widely, but typical trends may be highlighted by appropriate exposure trials such as a series carried out by Andrew and Donovan (1970). Consideration of the materials specified for government applications also allows us to illustrate the particular properties that determine the choice of a temporary protective. In reviewing the results an attempt has been made to relate them, the qualities of the materials and the user's need into a working perspective.

The temporary protectives shown in Table 13.2 form the major part of the range used by the UK government departments.

Results of a series of exposure tests are given in Figs. 13.1–13.6. The temporary protectives were applied to panels (150 mm × 100 mm) of steel, copper, brass, aluminium alloy and magnesium alloy. One set of panels was exposed under cover in a box with louvred sides for one year in a nominally rural atmosphere, although a nearby oil-fired boiler ensured a moderate level of sulphur dioxide. A second set was similarly exposed for one year within 150m of a surf beach at a location well removed from industrial pollution. A third set was subjected to a salt droplet test (BS 1391) for one month. A fourth set of panels was stored in a clean environmental chamber for five years in warm moist conditions (30°C, 85% RH). The conditions of exposure of the first set of panels may be likened to average conditions in an unheated storehouse in an urban area and the second to storage in a very corrosive environment — typically under cover on the deck of a ship; the conditions of exposure of the fourth set were in a clean environment and any corrosion would be caused either by contaminants already on the metal surface before the protective was applied or by an ingredient of the protective.

Three points of key interest are apparent from these results:

- the corrosion rate of unprotected steel is an order of magnitude greater than that of the other metals;
- the extent of protection afforded to steel by temporary protectives is mainly dependent on coating thickness;
- some temporary protectives cause corrosion of some non-ferrous metals.

In the warm damp atmosphere test, after five years most panels were unattacked. The exceptions were the extensive corrosion on copper, brass and magnesium in contact with PX-15 and to a lesser extent with PX-12. The probable causes are the lead stearate in PX-12 and the long-chain fatty acid ester plasticisers in PX-15. The bright green corrosion products which

Table 13.2 — Temporary protectives used by UK government departments

Type	Description	BS 1133 (Type)	Ingredients and field of use
(I) Oil type thin film	PX-4	TP6	An inhibited mineral oil. General purpose protective
	PX-14	TP2a	A mineral oil containing lanolin for the preservation of aluminium alloy sheets
	OM-17	TP6	A water-white, stabilised mineral oil, used on equipment to come into contact with food
	OX-18	TP6	A light inhibited oil, used for cleaning and short term protection; some water displacing properties
	PX-25	TP6	A light oil containing both contact and volatile corrosion inhibitors
	OX-275	TP6	Corrosion inhibited mineral oil for the internal protection of aircraft piston engines during storage (formulated to combat the corrosive effects of the combustion products of leaded fuels)
(II) Oil type thin film water displacing	PX-10		A thin inhibited liquid, used for preliminary treatment of salvaged mechanisms and components after alkali degreasing
	PX-24		A thin inhibited liquid, leaves a protective film, used for preliminary treatment of salvaged mechanisms and as a maintenance treatment on delicate or valuable items exposed to salt spray
(III) Grease type	PX-19	TP4a	A soft petroleum oil/calcium soap grease containing corrosion and oxidation inhibitors. General purpose use
	XG-250		A silicone grease for waterproofing, protecting and insulating ignition systems and electrical equipment
	PX-12	TP4b	A soft, castor oil/lead stearate grease used for the protection of light alloys or steel in assemblies which include rubbers in their construction. (Should not remain in contact with rubbers: not to be used on copper based alloys nor on cadmium coated surfaces)
	PX-13	TP4a	A wax thickened corrosion inhibited oil for the preservation of aircraft engines and Naval diesel engines. (Formulated to combat corrosive attacks of the combustion products of leaded fuels)
	PX-6	TP5	A stiff, rather tacky petrolatum
	PX-7	TP5	A soft petrolatum, used mainly for battery terminals
(IV) Soft film—cold application	PX-1	TP2a	A solvent deposited lanolin material for preservation at medium ambient temperatures
	P.A.2		A starch thickened 12%/w/w aqueous solution of sodium nitrite. For preservation of steel components stored unpackaged
(V) Soft film—hot application	PX-11	TP3	A petrolatum/beeswax protective for use under temperate and tropical conditions
(VI) Hard film—cold application	PX-2		A bitumenised material containing castor oil and zinc naphthenate. For the preservation of spares, tools, machined surfaces, etc. during storage. Not for internal surfaces, moving parts or inaccessible recesses (trichloroethylene solvent)
	Composition rust preventive Type B		A bitumenised material containing castor oil and zinc naphthenate. For the preservation of spares, tools, machined surfaces, etc. during storage. Not for internal surfaces, moving parts or inaccessible recesses (not containing chlorinated solvents)
	PX-3		A lanolin-resin, zinc chromate pigmented material for general long term preservation at medium and high ambient temperatures. Mainly used where removal is not essential
	PX-9	TP2a	A lanolin-resin protective suitable for general use
(VII) Strippable coatings	PX-15	TP7	A tough, transparent plastic material containing oil, inhibitors and plasticisers
(VIII) Wrapping papers impregnated with volatile corrosion inhibitors (VCIs)		TP9	Kraft paper (basis wt. 55 g/m^2) with corrosion inhibitor Type I (dicyclohexylamine nitrite, DCHN, 10-20 g/m^2)
		TP9	Kraft paper (basis wt 135 g/m^2) with corrosion inhibitor Type I (10-20 g/m^2)
		TP9	Kraft paper (basis wt 55 g/m^2) with corrosion inhibitor Type 2 (DCHN, 10-20 g/m^2 and cyclohexylamine carbonate, CHC, 10-20 g/m^2)
		TP9	Creped kraft paper (basis wt 85-95 g/m^2:) with corrosion inhibitor Type 2 (10-20 g/m^2 of both DCHN and CHC)

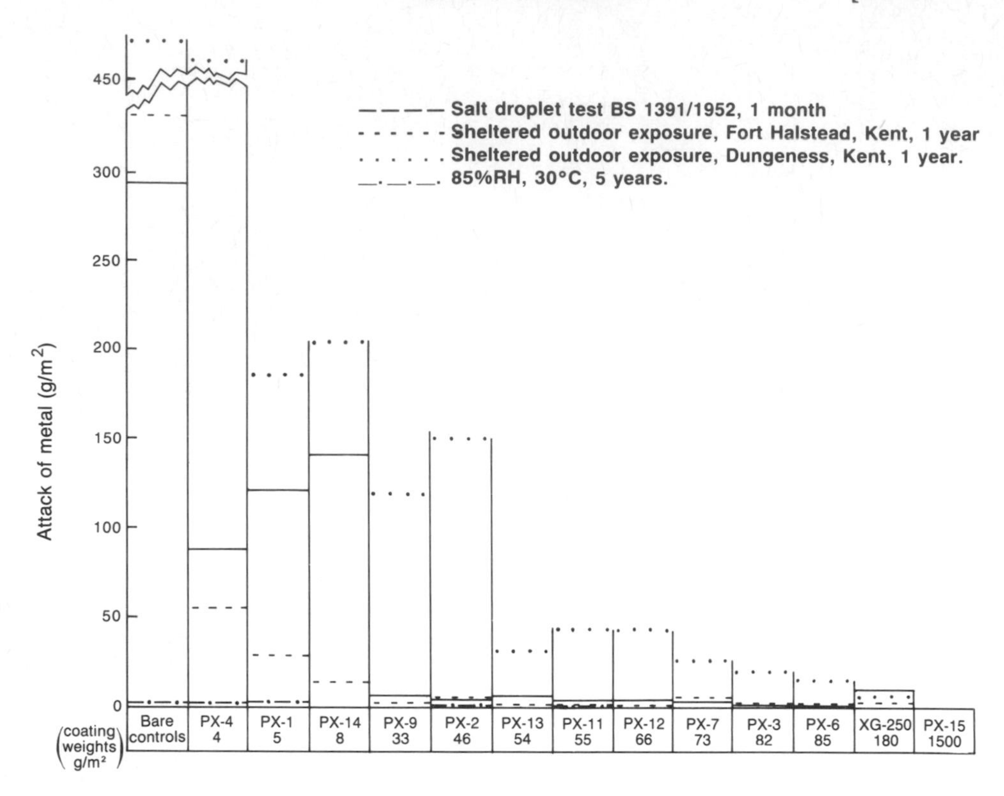

Fig. 13.1 — Protection of steel (BS 1449 Pt 1 HR1) with temporary protectives. (© Controller, Her Majesty's Stationery Office 1986.)

formed on the copper are typical of the products formed by reaction of long-chain fatty acids with copper. The unprotected panels were unattacked, confirming that the protectives were responsible for the corrosion that had occurred.

In the semi-rural atmosphere all the protectives with the exception of the oil PX-4 and possibly the other two thin coatings, PX-14 and PX-1, gave satisfactory protection to all five alloys. Under the more severely corrosive marine environment PX-4, PX-14, PX-1, PX-2 and PX-9 were unsatisfactory on steel, with PX-12 giving a marginal protection of steel and accelerating corrosion of copper alloys. Inspection of the coating weights shows reasonably good correlation with protective performance. Coating weights above 10 g/m^2 give satisfactory protection to steel for 12 months in the covered semi-rural environment while coating weights above about 50 g/m^2 were needed for protection over the same period and the same exposure conditions in the marine environment. The tests also showed that salts of fatty acids cause accelerated corrosion if they are present in protectives on copper and aluminium alloys, and similarly esters of long-chain fatty acids corrode copper alloys.

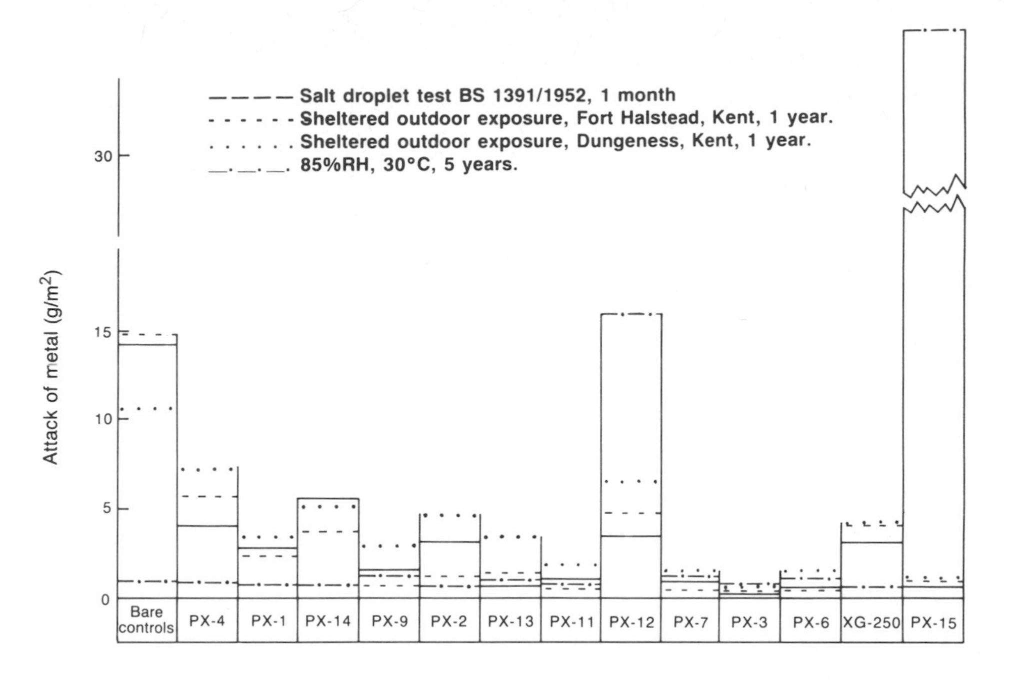

Fig. 13.2 — Protection of brass (BS 2870 CZ half-hard) with temporary protectives, see Fig. 13.1 for film weights.. (© Controller, Her Majesty's Stationery Office 1986.)

The relatively poor performance of the oil PX-4 on exposure for one year is to be expected since the role of oils within the spectrum of temporary protectives is for short-term protection on delicate mechanisms. Application of a protective oil is usually an integral part of cleaning and maintenance procedures on mechanisms with sliding and bearing surfaces such as guns, engine components, clocks and electro-mechanical components where surfaces are subjected to wear, stresses and high temperatures and yet are required to maintain close tolerances on both overall dimensions and surface finish. OX-18 is a typical light oil which contains a high level of contact corrosion inhibitors and has useful water-displacing properties so that it can be used after aqueous treatments and gives protection under mild conditions for several weeks; where water displacement is a major requirement the specially formulated PX-10 is preferred. PX-25 is particularly useful for the internal preservation of engines during storage and transport since the volatile inhibitor helps to provide continuous protection to the regions above the oil level. PX-24 is a representative of a class of material applied by spray aerosol which has become popular over recent years for protecting electrical assemblies and removing condensed water; it replaces the aqueous film on the surface with silicone fluid and thus decreases electrical leakage. It also loosens dirt and grease. It has been used in government service as a first defence on assemblies unexpectedly or necessarily subject to corrosive conditions — it has for instance been used as an

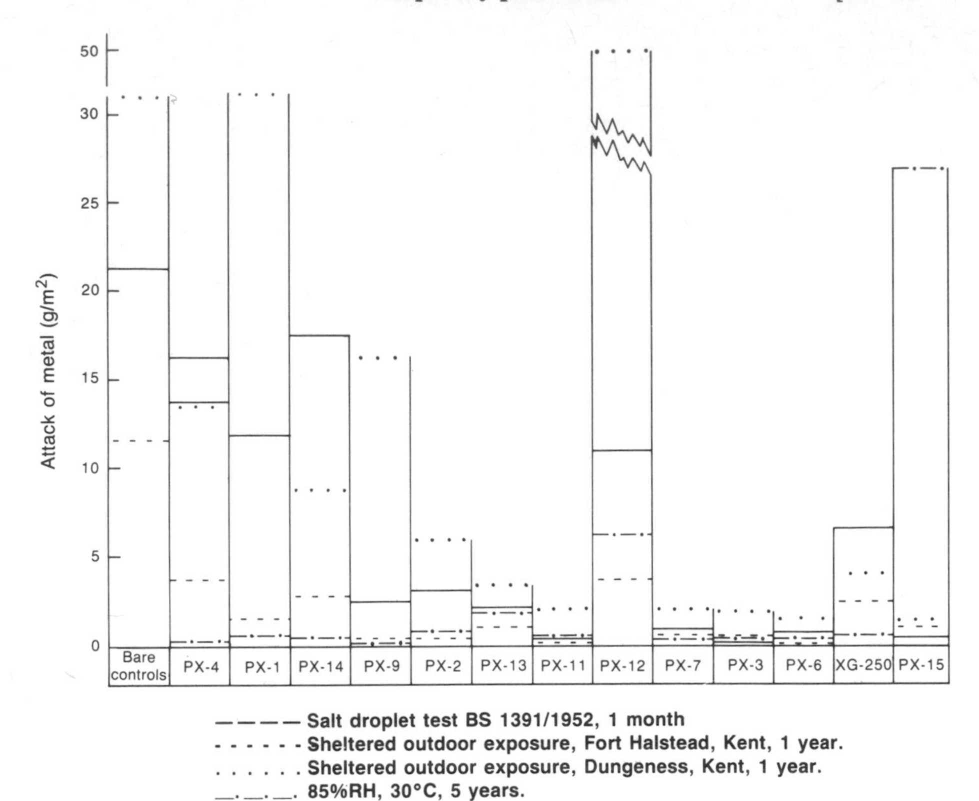

Fig. 13.3 — Protection of copper (BS 2870 C106 half-hard) with temporary protectives, see Fig. 13.1 for film weights.. (© Controller, Her Majesty's Stationery Office 1986.)

initial treatment on helicopters recovered from the sea and as a routine treatment on those regularly operated in marine environments. Each of the five oils listed has its well-established area of use and attempts to reduce the number have been unsuccessful. PX-25 for instance, although giving protection as good or better than OX-18 or PX-4, is not acceptable on exposed functioning mechanisms near personnel since spray contains the volatile inhibitor which can cause skin irritations especially near the eyes. OX-18 has been tried as a substitute for a thicker oil, Oil A (now replaced by PX-4), with what looked at first to be complete success as judged by early trials and salt droplet tests (see Fig. 13.7).

However, experience in use in warm climates soon revealed that protection from OX-18 was very short-lived. Tests on steel panels showed that the lighter oil both drained rapidly and also evaporated quickly; half the oil on a panel coated with OX-18 evaporated within four weeks, while a similar panel coated with the thicker oil lost only 5% of the oil coating in the same period (see Fig. 13.8).

Evaporation, drainage and the effects of dust deposits are the major

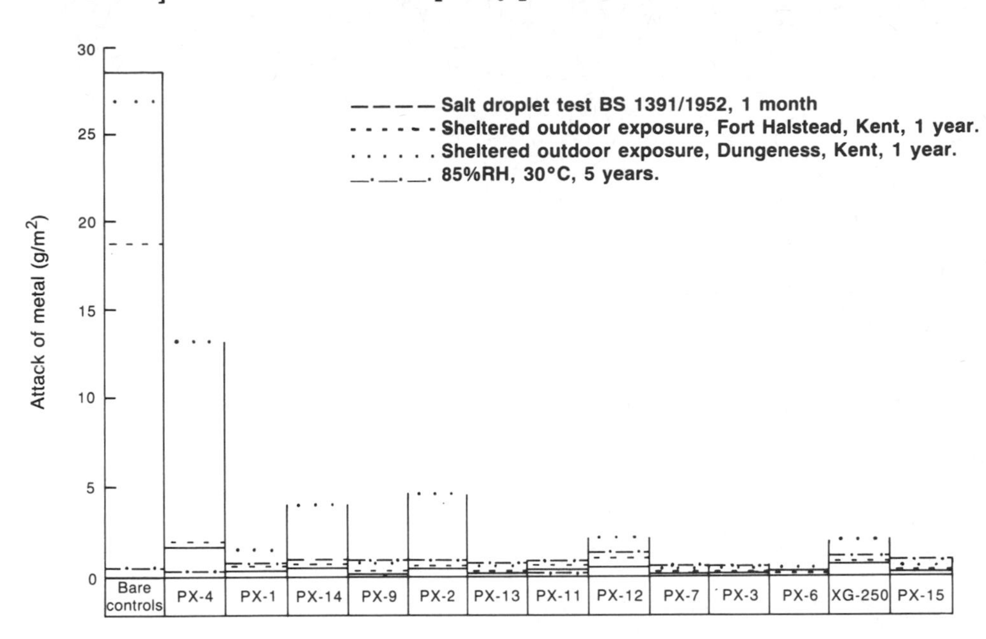

Fig. 13.4 — Protection of aluminium alloy (BS 1470 2014A T6) with temporary protectives, see Fig. 13.1 for film weights. (© Controller, Her Majesty's Stationery Office 1986.)

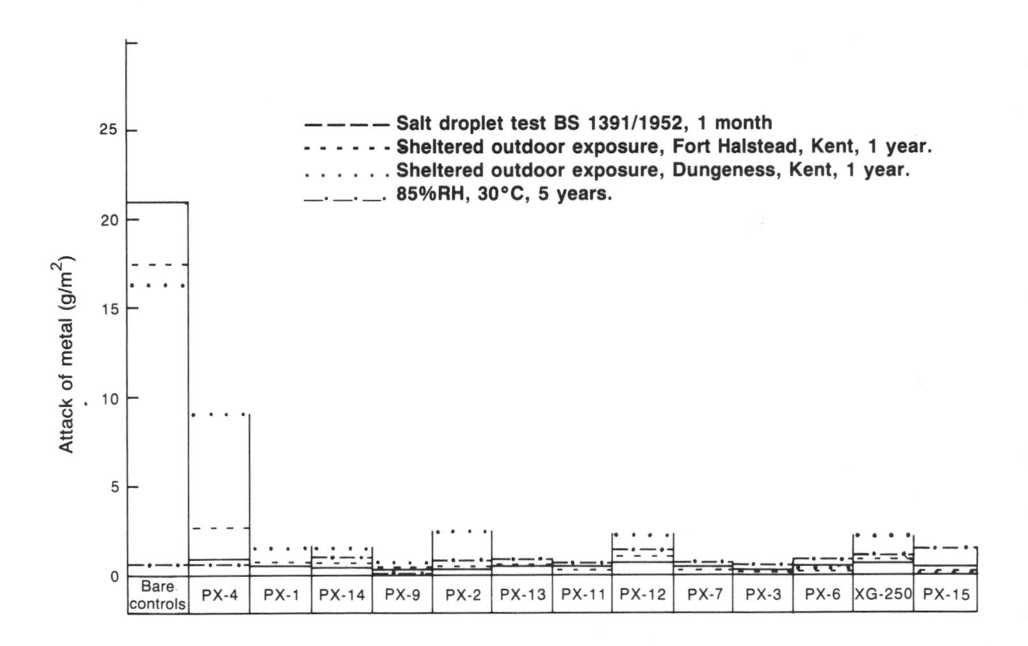

Fig. 13.5 — Protection of aluminium alloy (BS 1470 6082 T6) with temporary protectives, see Fig. 13.1 for film weights. (© Controller, Her Majesty's Stationery Office 1986.)

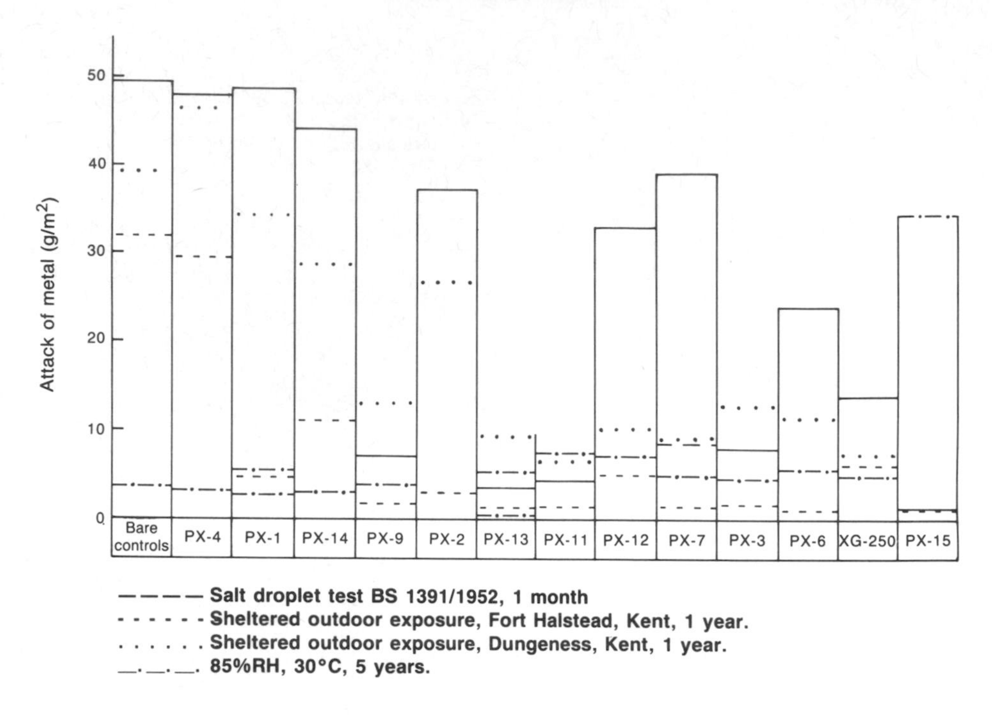

Fig. 13.6 — Protection of magnesium alloy (BS 2L 504) with temporary protectives, see Fig. 13.1 for film weights. (© Controller, Her Majesty's Stationery Office 1986.)

failure mechanisms that limit the effectiveness of oils, and these are not evaluated in short-term tests which are therefore often misleading for oil protectives.

INHIBITED WRAPPING PAPER: CONTACT INHIBITOR

Corrosion may be suppressed by wrapping components in paper impregnated with inhibitors. Fig. 13.9 shows for instance how sodium benzoate-impregnated paper reduces rusting by fingerprint residues.

Sodium benzoate impregnated papers give good protection to steels against tarnishing in the area of contact. Their use is largely restricted to simple components, particularly to sheet or strip. Results of some controlled tests are reported, in conjunction with those on volatile inhibitors, in the next section.

Cotton & Scholes (1970) have shown that benzotriazole-impregnated papers are very effective in protecting copper and brass. They recommend a tissue paper impregnated with 2% w/w benzotriazole for interleaving bright rolled copper and brass sheet in storage and in transport. They reported that this impregnation had prevented serious tarnishing whch previously occurred in the condensing conditions encountered as ships passed through the Panama Canal. In experiments they showed that copper wrapped in treated paper was unstained after being immersed in sea water for one hour and then

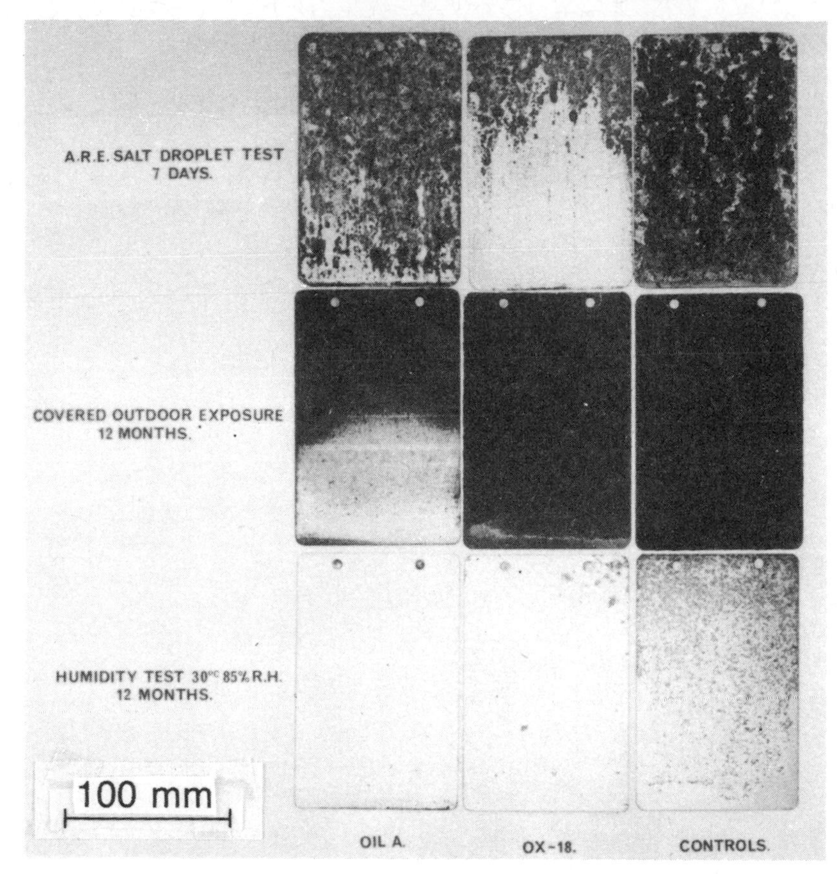

Fig. 13.7 — Steel panels treated with protective oils after corrosion test.
(© Controller, Her Majesty's Stationery Office 1986.)

allowed to dry, or after 24 hours' exposure to a 3% salt mist. In contrast, copper wrapped in untreated paper was severely stained after these tests. They reported similarly good performance in atmospheres containing sulphur dioxide, hydrogen sulphide and other pollutants, but protection by the inhibited paper was not apparently obtained in ammoniacal atmospheres. Impregnated paper is particularly useful for interleaving between metal sheets or between nesting items with smooth contours.

Scott & Skerrey (1970) reported that interleaving and wrapping papers impregnated with inhibitors such as sodium chromate or sodium metasilicate give added protection to aluminium and that the inhibitors counter the effect of harmful constitutents in the paper.

VOLATILE CORROSION INHIBITORS (VCI) IN WRAPPING PAPERS

Volatile corrosion inhibitors (VCI) extend the protection offered by impregnated wrapping papers to areas out of contact with the paper, so that overall protection may be obtained on complex shaped components where com-

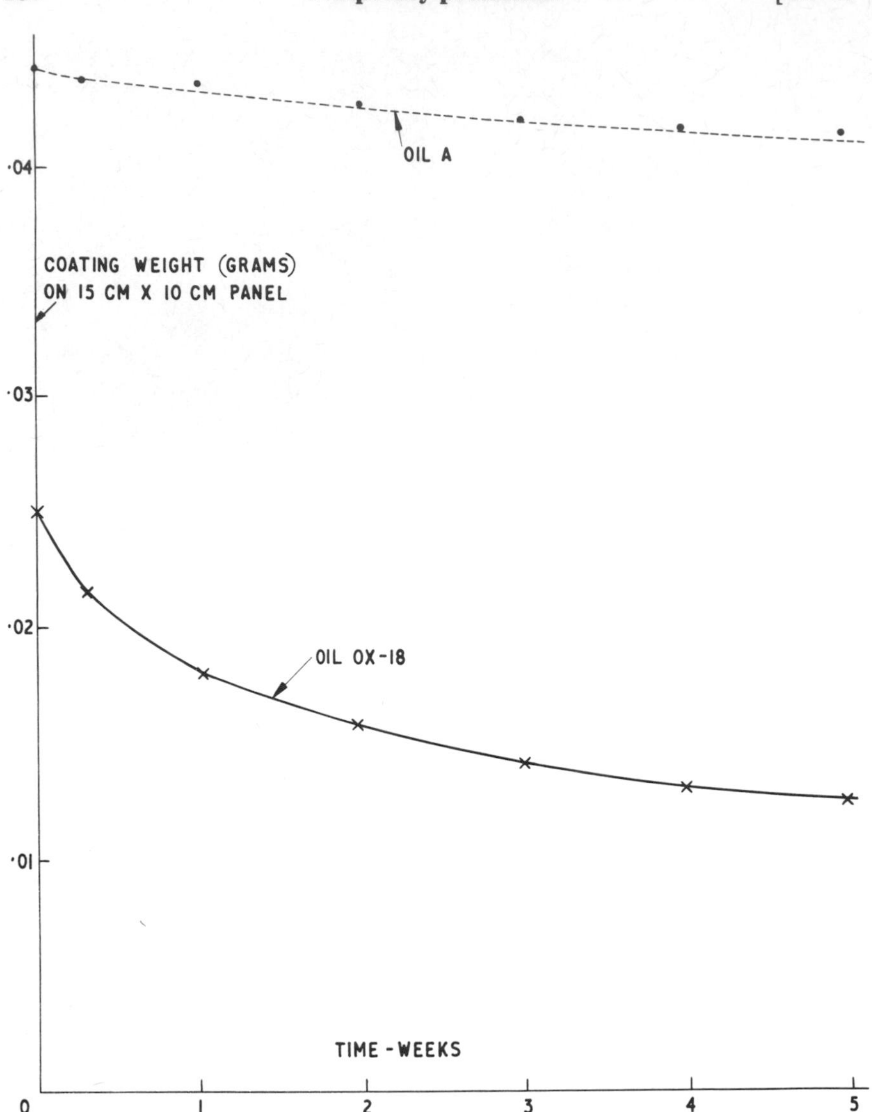

Fig. 13.8 — Weight losses of oils from evaporation. (© Controller, Her Majesty's Stationery Office 1986.)

plete contact with the wrap is not practicable. The VCIs which have found most favour are cyclohexylamine carbonate (CHC) and dicyclohexylamine nitrite (DCHN) which have vapour pressures of 0.4 mm and 0.0001 mm Hg respectively at ambient temperatures. The more volatile, CHC, gives a rapid build up of inhibitor in an enclosed space and penetrates further and more effectively than DCHN, but is more liable to become depleted, especially in poorly sealed packages.

Both are effective on steel but little information is available of their

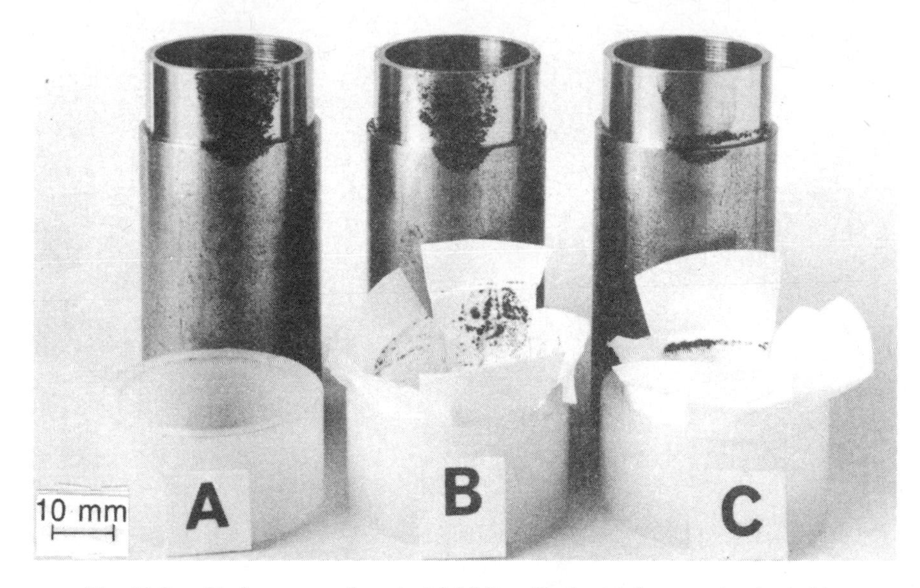

Fig. 13.9 — Performance of contact inhibitor. Steel containers contaminated by finger marking, after three week's storage. A, no paper in contact, steel corroded. B, unimpregnated paper in contact, steel corroded. C, paper containing 5% sodium benzoate in contact, corrosion of steel much reduced. (© Controller, Her Majesty's Stationery Office 1986.)

performance with non-ferrous metals. Available information indicates that DCHN may in some circumstances be aggresive towards magnesium, cadmium, zinc and lead, and CHC may attack copper and its alloys and magnesium. CHC is reputed to cause discolouration of some plastics. Bernie (1970) reported experience of VCI attacking coatings of paint and lacquer.

Experiments into the mechanism of protection by volatile organic amines indicate that the undissociated molecules of the inhibitor migrate to the metal surface where hydrolysis occurs. The amines, and probably the nitrite of the DCHN, act as anodic inhibitors, but they may also act in some environments by absorbing acid vapours directly and by maintaining the pH within the alkaline range on the metal surface.

Figs. 13.10–13.15 show some results of an extensive trial carried out by the writer and his colleagues on VCI wrapping papers to establish their value in protecting packaged bare steel components as an alternative to oils or greases. In these trials two types of test component were used, simple and complex, as shown in Figs. 13.10 and 13.11.

A range of inhibited wraps were evaluated including kraft paper containing 15 g/m^2 of DCHN, CHC, and a mixture of DCHN and CHC. Two proprietary products containing VCIs designated wrap A and wrap B were included for evaluation. Paper treated with the contact inhibitor, sodium benzoate, and untreated paper were included as controls. The base papers were tested and shown to comply with the requirements of DEF STAN 13–16 (paper, pure, kraft) and with the corrosion of steel plate test of DEF–1316.

COMPONENT 'A' WRAPPED IN DCHN/CHC KRAFT. RATING 0.
COMPONENT 'B' WRAPPED IN SODIUM BENZOATE KRAFT. RATING 8
COMPONENT 'C' WRAPPED IN PURE KRAFT. RATING 10.
EACH COMPONENT FINALLY CUSHIONED WITH CORRUGATED CARDBOARD AND PLACED IN CARDBOARD CARTON.

Fig. 13.10 — Simple components after five years of storage in the Far East.
(© Controller, Her Majesty's Stationery Office 1986.)

Four methods of packaging were used with both the simple and the complex components viz.

- wrapped in test kraft;
- wrapped in test kraft, waxed;
- wrapped in test kraft, then wrapped in bitumenised kraft;
- wrapped in test kraft, placed in a polythene envelope and heat sealed.

After wrapping, each component was cushioned in corrugated cardboard and placed in a cardboard carton.

Packages were stored in covered ambient temperate conditions (UK), covered ambient tropical conditions (Singapore), in a clean environment at

COMPONENT 'A' WRAPPED IN DCHN CREPED KRAFT FOLLOWED BY BITUMENISED KRAFT. RATING 2.

COMPONENT 'B' WRAPPED IN DCHN CREPED KRAFT AND HEAT SEALED IN POLYTHENE ENVELOPE RATING 8.

COMPONENT 'C' WRAPPED IN CREPED KRAFT AND HEAT SEALED IN POLYTHENE ENVELOPE. RATING 10.

EACH COMPONENT FINALLY CUSHIONED WITH CORRUGATED CARDBOARD AND PLACED IN CARDBOARD CARTON.

Fig. 13.11 — Complex components after five years of storage in the Far East. (© Controller, Her Majesty's Stationery Office 1986.)

30°C and 85% RH, and under DEF STAN 07-55 (Part 2) ISAT(A) environmental test conditions (see Chapter 2 for cycle).

After storage, components were inspected and marked according to the extent of increasing rust on a scale of 0–10. Typical examples are shown in Figs. 13.10 and 13.11; components rated 1 or 2 would probably be acceptable for most purposes. The results of these tests are summarised in Figs. 13.12–13.15.

The trial confirmed the effectiveness of DCHN and CHC together (total 15 g/m^2), impregnated in kraft wrapping paper, and DCHN alone (15 g/m^2), in protecting both simple and complex components for up to five years in temperate conditions. Protection was unreliable in the warm humid tropical environment. The conditions established for good protection in this and previous tests were: a high standard of cleaning of the metal surfaces

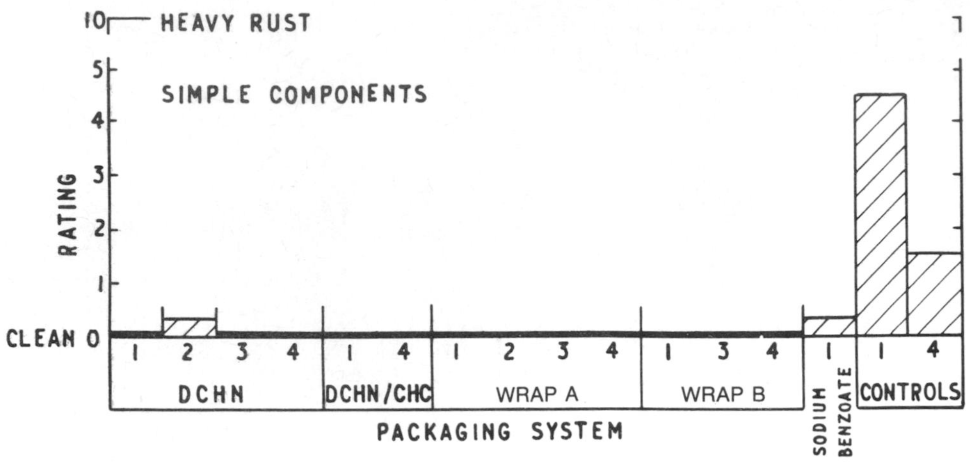

PACKAGING SYSTEMS (ALSO FOR FIGS. 13.13–13.15)

1 WRAPPED IN VCI TREATED KRAFT

2 WRAPPED IN VCI TREATED CREPED KRAFT, WAXED

3 WRAPPED IN VCI TREATED KRAFT THEN WRAPPED IN BITUMENISED KRAFT

4 WRAPPED IN VCI TREATED KRAFT, PLACED IN A POLYTHENE ENVELOPE AND HEAT SEALED

AFTER WRAPPING AS IN 1-4 EACH COMPONENT WAS CUSHIONED IN CORRUGATED CARDBOARD AND PLACED IN A CARDBOARD CARTON

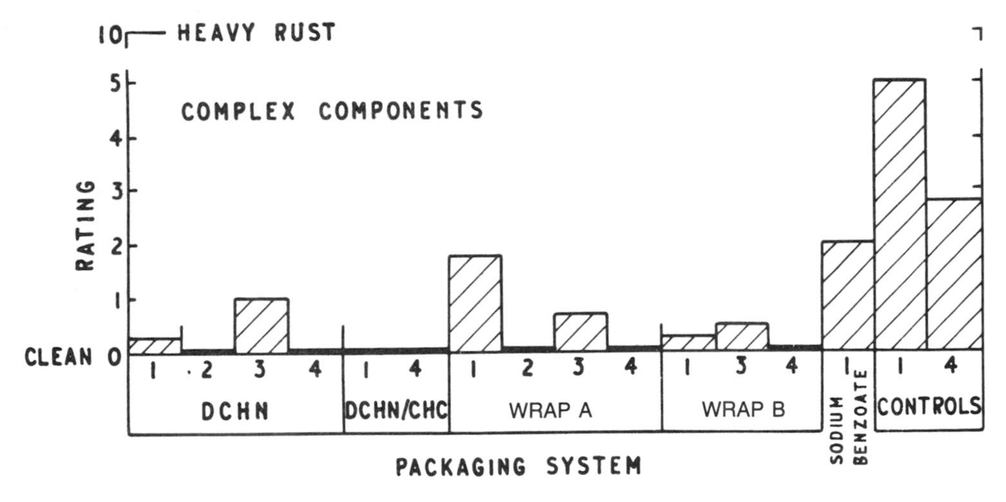

Fig. 13.12 — Protection of steel components with VCIs: temperate storage (five years). (© Controller, Her Majesty's Stationery Office 1986.)

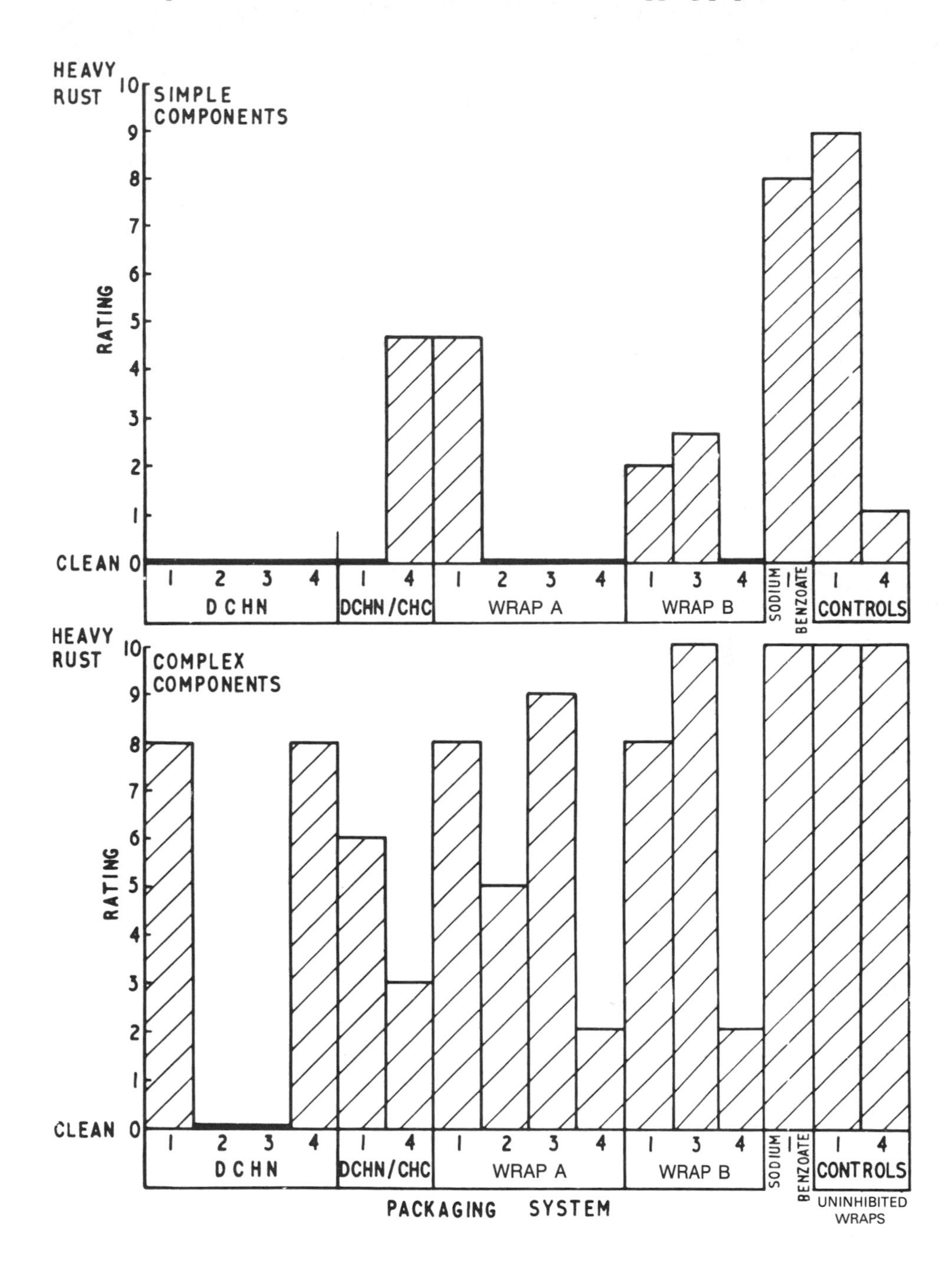

Fig. 13.13 — Protection of steel components with VCIs: tropical storage (five years). (© Controller, Her Majesty's Stationery Office 1986.)

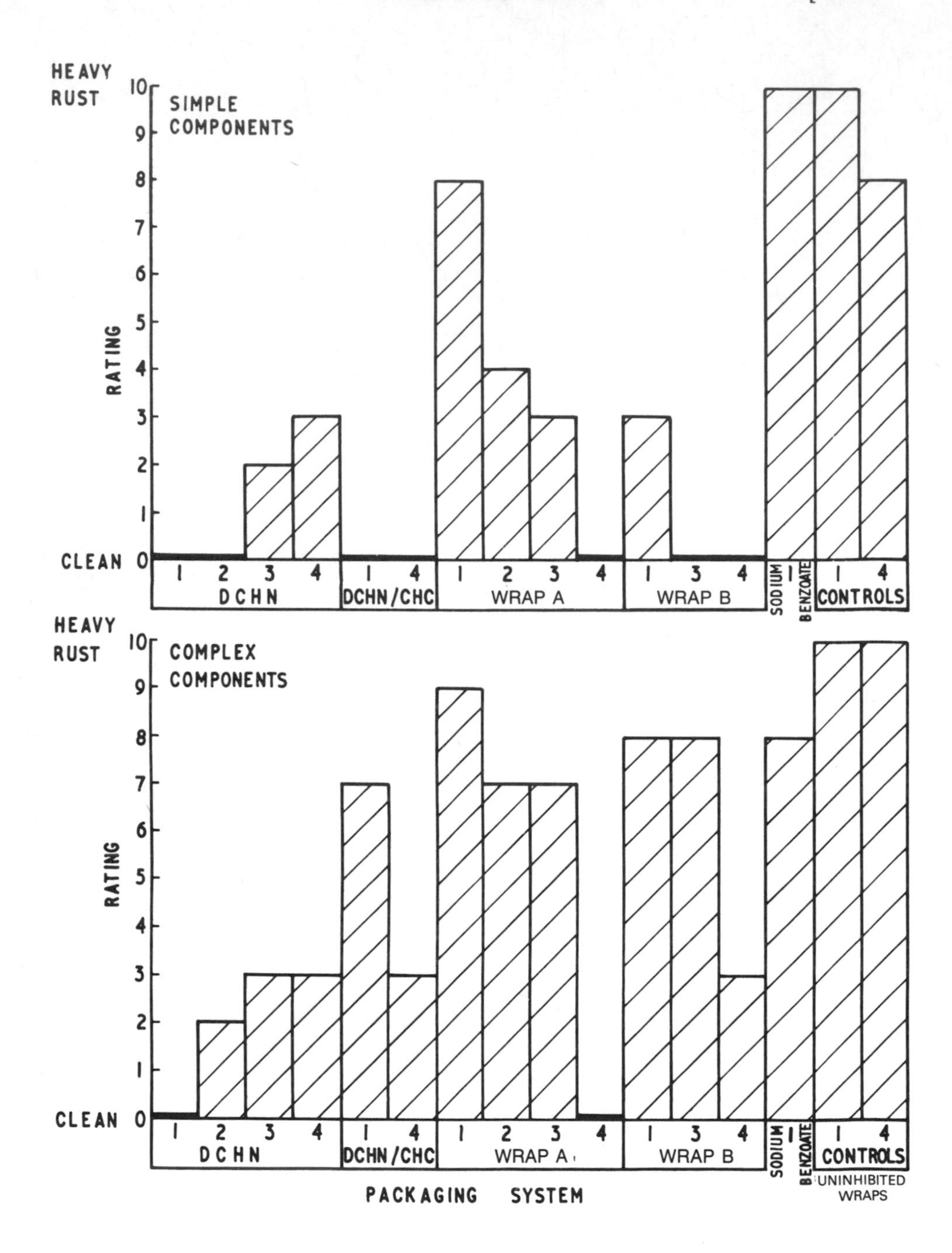

Fig. 13.14 — Protection of steel components with VCIs: 30°C and 85% RH (one year). (© Controller, Her Majesty's Stationery Office 1986.)

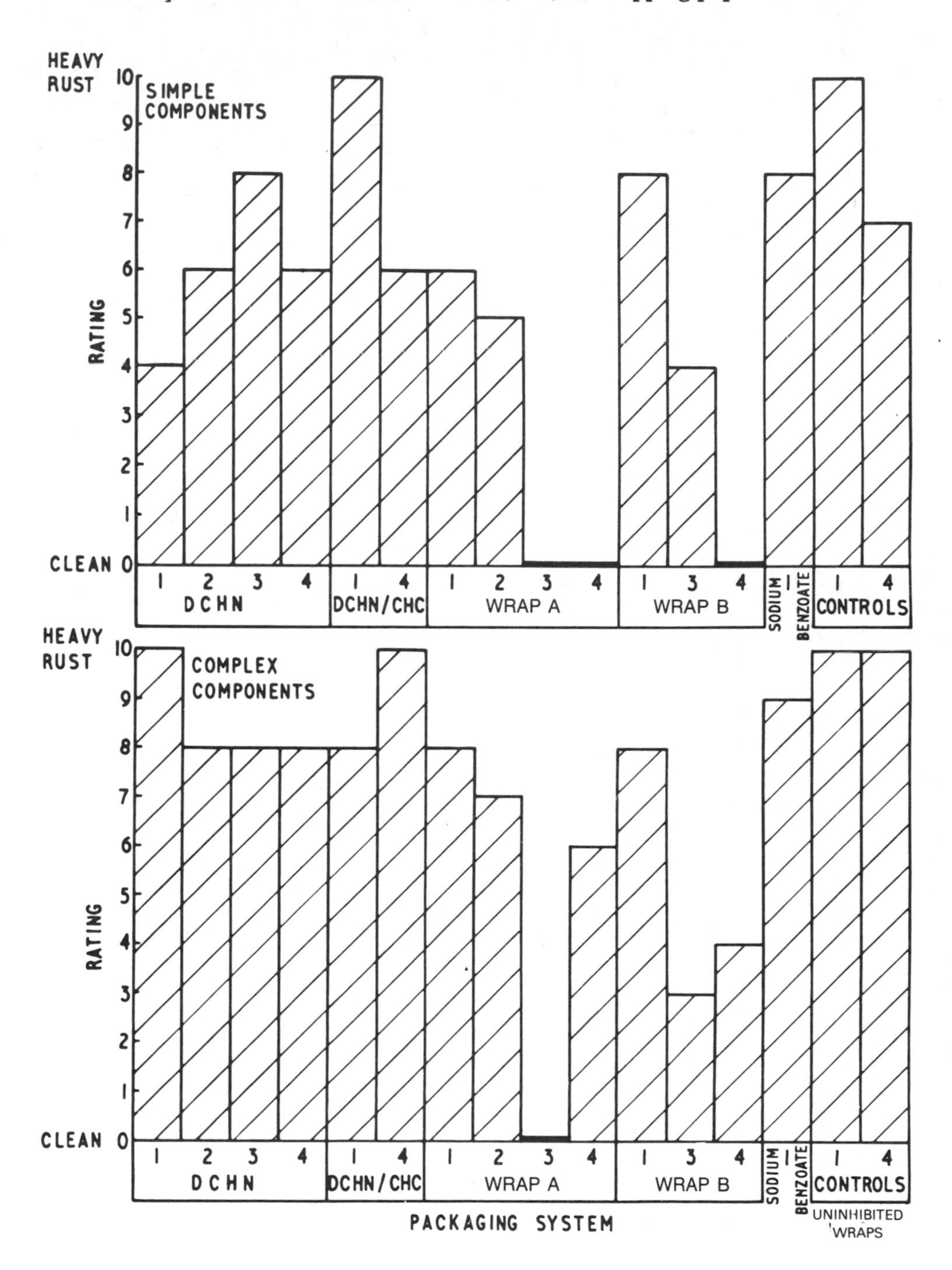

Fig. 13.15 — Protection of steel components with VCIs: ISAT(A) (one year). (© Controller, Her Majesty's Stationery Office 1986.)

followed by rapid and careful packaging; an impervious overwrap of either polythene or bitumenised kraft, and an absence of corrosive vapours. The need for this last was inadvertently demonstrated by unexpected corrosion with preliminary tests on one manufacturer's inhibited papers which investigation showed was caused by acetic acid released by hydrolysis of polyvinyl acetate used as a binder in that product.

The comparatively heavy rusting of the controls in the warm damp atmosphere indicated that either impurities in the paper, or residues from the trichloroethane degreasing, were causing some corrosion.

VCIs may also be introduced as either a loose powder in a porous bag or in a 5% solution in neutral methylated spirits which is sprayed on to the surfaces needing protection. With VCIs it must be emphasised that, however they are applied, an overwrap or external package is needed to retain the vapour of the inhibitor. Although they afford protection under temperate storage conditions, they should not be relied upon for long-term protection in more adverse conditions when film-forming temporary protectives should be preferred.

Volatile inhibitors are suited to protecting delicate components such as precision tools, gauges and instruments where protection by oils and greases would be inappropriate; added to oils used to give protection within enclosures they extend protection to surfaces which the oil does not readily reach or from which it drains. In appropriate conditions VCIs have a useful role to perform but their reputation has in the past been tarnished by extravagant claims and by use in unsealed or too corrosive environments; the basic concepts of clean surfaces, uncontaminated atmospheres, a well-sealed enclosure, limited distances from inhibitor to metal surfaces and adequate quantities must again be emphasised as fundamental requirements if VCIs are to give satisfaction.

REMOVAL OF TEMPORARY PROTECTIVES

Temporary protectives can be removed, where this is necessary, with solvents, either with white spirit applied by brush or swab, or by immersion, or with halogenated hydrocarbons in an enclosed degreasing unit. Some of the wax-based protectives, and especially lanolin, harden and cross-link with long-term ageing and may then be difficult to remove with solvents and some abrasion may be needed in conjunction with the solvent. Oil- and grease-based protectives are often compatible with the in-use environment and do not need to be removed, or it may be sufficient to remove the bulk with a clean dry cloth. Strippable coatings peel off but the remaining film of oil should not be forgotten — although its presence will often be an advantage for the protection it confers, sometimes it may be necessary to remove it.

VCIs leave the surface essentially unaffected but through their mode of action they leave the surface oxide reinforced and some inhibitor will remain absorbed on the metal surface, especially at the areas where reactions would normally commence. If such items are to be plated or phosphated some reactivation of the surface will be needed.

EFFECTS OF STORAGE CONDITIONS

The selection of temporary protectives has to be based on the nature and complexity of the items to be protected, their value and most importantly the storage conditions and the duration of storage.

Environmental conditions are conveniently considered in three categories:

(a) Indoors.
Moderate humidities, protected from high levels of contamination and not subjected to the extremes of the external temperature (some further subdivision into heated and unheated stores is sometimes useful).
(b) Sheltered outdoors.
Variable humidity, wide range of temperatures and water condensation, but protection from rain and the worst effects of the weather.
(c) Unsheltered outdoors.
Full range of conditions — rain, temperature extremes, solar heating, marine spray, condensation and dust, sand and other particulate contaminants.

The severity of the environments depends on the ambient climate. Any packaging will modify these conditions by reducing the extremes of temperature and humidity reached, keeping out most environmental contaminants and giving physical protection. The thin (TP1 or TP2) or soft (TP3, TP4 or TP5) protectives, preferably with an overwrap, may be appropriate.

SELECTING A TEMPORARY PROTECTIVE

In selecting a temporary protective consideration must be given to the packaging, protectives already present, other nearby materials, the type of metal and its sensitivity to corrosion, the level of protection needed, costs, ease of application or removal, hazards in application (e.g. from fire), the time over which protection is required, the physical environment and consequent hazards and the severity of environmental conditions expected. Particular hazards to be established may arise from: humidity, temperature, condensation, vibration and rough handling, transport, the need for maintenance or inspection, attack by animals or mould, and ingress of heavy concentrations of corrosive vapours or marine spray. Most of these effects are dealt with in other parts of the book but they are all particularly important in limiting the life and effectiveness of temporary protectives. All but the most robust (TP7) may be seriously damaged by handling or by full exposure. Solar heating will cause softening and many greases then emulsify in subsequent condensation or rain (there are bitumen-based and grease-based paints formulated to withstand full exposure but these are not readily removed and are beyond the accepted definition of temporary protectives). Temporary protectives can only protect for a short period against full exposure and a tarpaulin cover or other protection from sun and rain should be provided; with this basic degree of protection the results of the marine exposure in Figs. 13.1–13.6 represent the levels of protection to be expected in severe conditions. Only tropical marine or deck cargo covered exposure would be substantially worse.

The physical properties of temporary protectives are an important consideration; soft films are readily damaged. Fig. 13.16, for instance, shows lines of corrosion which included extensive pitting, on rocket motor tubes protected with lanolin (PX-1). The lines were at 60° intervals around the periphery, marking the contact between tubes which had been stacked under cover. The rusting reduced the bursting strength of the tubes by 30%.

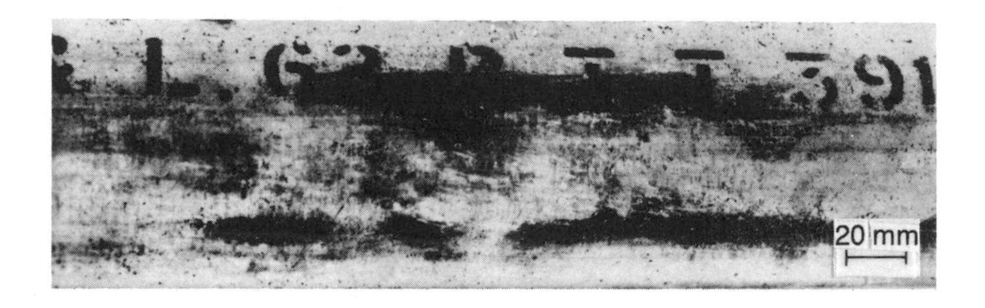

Fig. 13.16 — Rusting at contact lines between rocket tubes protected with lanolin (PX 1). (© Controller, Her Majesty's Stationery Office 1986.)

The rocket tubes were subsequently given supplementary protection with a greased paper overwrap.

During the processes of manufacture, short-term protection is often given by light oils, some of which are formulated with additions of solvent and surfactants to give water-displacing and cleaning properties; these are useful in processing, in removing residues of machining fluids, in removing sweat residues and other corrosive contaminants and protecting from tarnishing for a short period until a more substantial protective is applied.

A single temporary protective is usually adequate for simple components such as nuts and bolts but more complex assemblies may require two or more. The sharp edges of tools for instance are often protected by hot dipping into TP7 while the body is protected by thinner and less costly films of TP1 or TP2. The internal areas of engines, gearboxes, pumps and compressors may be conveniently protected with oils preferably with additions of VCIs while the exterior is protected by grease; in protecting the external surfaces of complex items the ends of small bore tubes and deep cavities should first be plugged with waxed paper or plastic plugs (not wooden plugs). The threaded ends of pipes are best protected with TP3 and covered with a plastic cap. Large enclosed volumes such as inside pipes, cylinders, gun barrels and boilers may be protected with VCIs impregnated into paper which must approach within 50–80 mm of all the surfaces to be protected. Clocks, watches and cameras and similar delicate mechanisms may be protected with VCIs but a well-sealed impervious external wrap or box is then needed; while both steel and copper alloy surfaces are readily protected in this manner, zinc, cadmium, aluminium and magnesium are not protected by VCIs and may be corroded by them.

COMMERCIAL PRODUCTS

There are several commercial ranges of temporary protectives available. Most have a selection across the range of types specified in BS 1133, although a recent sample survey by the writer of five product ranges produced only two for which the supplier was willing to quote the BS 1133 classifications. Each manufacturer offered a comprehensive range of products with technical descriptions of their physical properties and fields of use. Bryde (1970) gives a description of one proprietary series which does give the BS 1133 categories. He describes in informative detail the selection of appropriate products for protecting an extensive range of plant and machinery in storage and transit.

The protection offered by temporary protectives depends on the coating thickness and a close indication of the performance of a commercial protective on steel can be obtained consulting Fig. 13.1 for a product of similar thickness. Few manufacturers give information on the performance of their products on non-ferrous metals and possible incompatibility should always be considered.

REFERENCES

Andrew, J. F. & Donovan, P. D. (1970) *Proc. Conf. Protection of Metal in Storage and Transit,* London, Brintex Exhibitions, p. 25.

Beale, E. W. (1976) Shreir, L. L. (Ed.) *Corrosion.* Newnes-Butterworths, pp. 17, 21.

Bernie, J. A. (1970) *Proc. Conf. Protection of Metal in Storage and Transit,* London, Brintex Exhibitions, p. 43.

Bryde, R. (1970) *Proc. Conf. Protection of Metal in Storage and Transit,* London, Brintex Exhibitions p. 9.

Cotton, J. B. & Scholes, I. R. (1970) *Proc. Conf. Protection of Metal in Storage and Transit,* London, Brintex Exhibitions,p. 39.

Scott, D. J. & Skerrey, E. W. (1970) *Proc. Conf. Protection of Metal in Storage and Transit,* London, Brintex Exhibitions, p. 45.

14

Controlling humidity

Atmospheric corrosion can be prevented by maintaining the relative humidity below the critical humidity for corrosion. For most of the normally encountered metal/contaminant combinations, with the notable exception of marine salt, the critical humidity lies within the range of 60–70% (for marine salt Evans & Taylor (1974) showed that it was below 35%). Maintaining the humidity within enclosures to below 50% not only controls corrosion but also eliminates mould growth, rotting and the activity of mites, insects, termites and animals (at least on the inside). The highest level of control is attained by ensuring that the contents are initially dry and then keeping them dry by either fully sealing the enclosure or by using a combination of barriers and desiccants.

The need is greatest in damp tropical regions where the high vapour pressure of water ensures that large quantities of water are present in the air even at moderate levels of relative humidity, and variations of temperature may give copious condensation.

RELATIVE HUMIDITY (DEFINITIONS)

Definitions of the relative humidity and related terms are:

(1) *Relative humidity (RH)*. The ratio, expressed as a percentage, of the amount of moisture in the air to that in saturated air at the same temperature.

(2) *Vapour pressure.* That part of the atmospheric pressure created by water vapour.
(3) *Saturated vapour pressure.* The maximum partial pressure that can be exerted by water vapour at a given temperature.
(4) *Dew point.* The temperature at which a plane clean metal surface is just cold enough for visible condensation to occur on its surface.
(5) *Critical humidity.* The minimum humidity at which corrosion occurs with a given combination of metal and contaminant (the critical humidity is nearly constant over a wide range of temperatures).

SOURCE OF WATER IN ENCLOSURES

Free water is produced as precipitation from rain, snow, hail or fog; splash and spray, especially near the sea; from condensation on cold surfaces, and by migration from absorbent materials.

Providing cover against rain, snow, hail, splash and spray is a first step in protection. Materials can be adequately dried before being brought into storage areas or incorporated into packages. Penetration of air into enclosures and condensation of water on cold surfaces within them results from a complex series of processes and its suppression requires careful control of the materials used and of the environment.

Water will enter a partially sealed container by the processes of capillary action, diffusion, and by 'breathing' induced by both barometric pressure changes and temperature changes of the enclosed air.

Capillary action provides a mechanism for rapid penetration of any external liquid water through imperfections in seals, cracks, or the sleeving over electrical wiring.

The partial pressure of water vapour within an enclosure is generally less than it is in the outside air, so that diffusion leads to an increase of water inside the enclosure. It should be noted that a positive overpressure within an enclosure does not affect diffusion of water vapour which depends only on the difference between the partial pressure of water vapour on the inside and outside of the enclosure.

The first barrier to water vapour is provided by the external container. If this is of metal it may be made impervious to water and water vapour. But water vapour diffuses readily through most wood, cardboard and other natural products. Main structures made of plastics, such as glass-reinforced resins or ABS, are usually of such a thickness that water transmission rates are very low but not insignificant. Diffusion through thin plastic barriers is more rapid and its rate is often of great importance in packaging. Thus for example 18 g of water will diffuse per day through each m^2 of low density polyethylene of 25 μm thickness at 38°C when one side is maintained at 90% RH and the other below 2% RH.

Although diffusion through the tortuous leakage paths through seals and cracks is slight, a much greater quantity of water is drawn into enclosures through leakage paths by 'breathing', as a consequence of pressure and temperature changes. Fluctuations of barometric pressure give a continuing but slow interchange of air. In air freight, pressure changes are high and relief valves are usually fitted. In most other circumstances the effects of pressure changes are much less than those of temperature variations, which are usually the major cause of 'breathing', and the consequent transfer of water to the inside of enclosures. If the contents have a high heat content their temperature lag over the diurnal cycle leads to retention of condensed water within the enclosure. The effect is exacerbated by exposure to the sun, and is most serious in humid tropical conditions.

HUMIDITY AND TEMPERATURE

The variation of the saturated vapour pressure (100% RH) over the temperature range 0–70°C is shown in Fig. 14.1. From 0 to 40°C the water

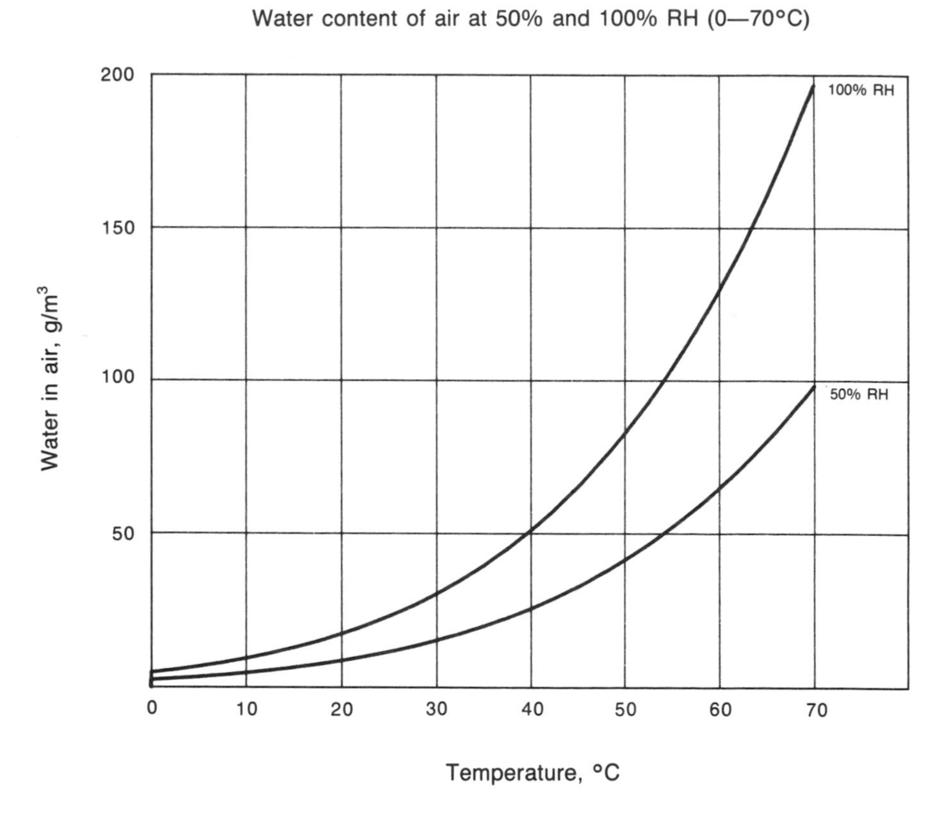

Fig. 14.1 — Water content of air at 50% and 100% RH (0–70°C).

content of saturated air doubles with each 11°C rise in temperature. Thus air at 50% RH, a safe level well below the critical humidity for most metal/ contaminant combinations with the notable exception of steel with marine salt, cooled by 11°C becomes saturated with water vapour and metals present are liable to corrode. Conversely the relative humidity of saturated air, heated through 11°C, falls to 50% RH and corrosion practically ceases.

The effects of temperature changes on the corrosion of metals in enclosures are however often more complex than the simple relationship of temperature with relative humidity might suggest. This is because the heat lag of goods of high heat capacity and the hysteresis induced in the relative humidity by water retained by materials present, modify the conditions on the surface of metal items in enclosures which are subject to temperature changes.

A typical incident reported by Middlehurst & Kefford (1970) illustrates some of these effects. The authors encountered corrosion and disfiguring staining of canned foodstuffs which occurred intermittently in a series of shipments from Australia to Europe. The goods were stored in well-ventilated holds and the damage was most pronounced at the centre of the stacks. Temperature monitoring showed that the cans, which were loaded under temperate conditions, were much cooler than the warm ambient air as the ship passed through the topics; condensation occurred as the air circulated and was cooled to below its dew point. The high heat capacity of the goods resulted in a prolonged temperature lag and condensing conditions were maintained over a protracted period with consequent staining and corrosion. The problem only occurred when the cans were stowed during the winter months when their temperature averaged about 7°C. Twenty to ninety per cent of cans in affected shipments showed marked staining. The effects were more serious if absorbent materials such as wool were stored within the same hold. Corrosion was reduced by discontinuing the circulation of ambient air and surrounding the stack with kraft/polythene or kraft/bitumen wraps. Staining did not occur when the cans were transported in sealed containers although occasionally staining was still encountered when the doors of a container had been sprung open by rough handling during loading.

Condensation may be expected when goods are subject to diurnal temperature changes, are moved from cool exterior storage into heated buildings, or are subjected to climatic changes during transport; the contrasts in air freighting are particularly pronounced since goods are often exposed in flight to low pressures and sub-zero temperatures and may be off-loaded minutes later to be stowed in containers or storehouses in which warm damp air circulates and condensation readily occurs.

Water collects in incompletely sealed containers especially in tropical and semi-tropical conditions of high humidity because of the large swing of ambient temperatures with the diurnal cycle, and the heat lag of the contents of the container. It is often better to insert drain holes to ensure that water does not accumulate rather than risk this build-up of condensation in an incompletely sealed container. A flexible inner barrier helps in overcoming the effect; enclosed absorbent materials, and exposure to direct sun increase the risk of water accumulating.

Natural materials retain a high proportion of water in a loosely absorbed state, the amount retained increasing with the relative humidity. If such materials are introduced into packages or enclosures they must be dry or high humidities will result.

Wood in the as-felled or green condition has a moisture content of 50–80%. When the wood has been dried to equilibrium with air at a relative humidity of 50% its moisture content falls to 15–18%. Hair, which is sometimes used bonded with latex as a cushioning, felt, wool, cloth, jute and paper retain 20–25% of water in equilibrium with air at 90–95% RH which reduces to 5–10% in equilibrium with air at 50% RH. Synthetic organic materials retain much less water although 1–2% of their weight may usually be removed by drying from equilibrium at 95% to 50% RH. Dried absorbent materials exposed to high humidities will re-absorb water.

If wood which has been stored under conditions of high humidity is used, without subsequent drying, in fabricating packages or their internal furniture then large quantities of water may be introduced into the package. Thus a package of one cubic metre capacity which contains 5% of its volume as wood could contain over 1% of its volume as readily available liquid water, i.e. 10 kg of water, absorbed within the wood if the timber were taken directly from an outside store in which the relative humidity was 80–90%. The surplus water is difficult to remove and would maintain high levels of relative humidity within the package. In contrast, ambient air at 30°C and 100% relative humidity filling the whole of the package would only introduce 0.03 kg of water. If natural materials are introduced in quantity into any enclosure they will have a dominant effect on the internal humidity and it is important that only dry materials are used if metals are to be present. A possible alternative is to purge the container with dry air until the absorbed water is removed.

HUMIDITY DURING PACKAGING

Reducing the relative humidity within packages to as low a level as is practicable is a key requirement in preventing corrosion (and incidentally for suppressing mould growth and most other types of biological activity). This requires that the contents are initially dry, that water is excluded by sealed waterproof or water-vapour-proof barriers, and, for the highest standard of corrosion suppression, that desiccants are incorporated. It is also important that equally dry conditions are attained whenever packages are opened and resealed.

The relative humidity within the storage area should be monitored and where possible the dew point of both the incoming and outgoing air should also be monitored. In some climatic conditions, during part of the year — typically summertime in temperate areas, the ambient air is sufficiently dry and conditions can be controlled by the rate of air intake, but in the cooler months adequately dry storage is more readily maintained in large storage areas in temperate climates by heating the atmosphere to about 10°C above the mean external temperature, supplemented as necessary by trays of desiccant. In warm humid climates it is often preferable to dry air by refrigeration. The humidity of the storage area should be at the lowest level at which the materials maintain their desirable physical properties and dimensional stability. Paper for instance becomes brittle when very dry. A relative humidity of about 50% is appropriate for the storage of most materials used in packaging, but individual specifications should be consulted since optimum storage conditions vary slightly with the type of materials.

The difference between the water vapour content and relative humidity, and the sensitivity of the latter, but not the former, to temperature changes, is not always appreciated. An experience at an instrument repair shop once impressed the writer with the ease of misunderstanding on this subject. A young Flight Lieutenant explained proudly that after repair, an instrument was placed in a polythene bag which he then purged with warm air from a hair dryer, and quickly sealed. That ensured that the air inside was thoroughly dry, didn't it? The consequent long discussion at least left a much clearer impression with one of us of the need to be clear on the distinction.

SEALED PACKAGES

Vacuum packing, or sealing under dry nitrogen, gives long-term protection if the contents are first dried, provided that the container is made of impervious material and is sealed. The only fully impervious seals are soldered, welded or brazed metal, or fused glass, although well-made compression joints using indium or gold gaskets have negligible leakage rates for most practical circumstances. Typical rubber or plastic compression joints, when well made and new, have water vapour leakage rates, at 38°C and 90% differential, of about $3\text{–}10\times10^{-3}\ \mathrm{gm^{-1}d^{-1}}$, but age, resealing or poor assembly give higher rates.

With adequate seals, vacuum packing or sealing under dry nitrogen has the added advantage over desiccation of excluding not only water vapour but a second essential of corrosion, atmospheric oxygen, but any inspection of the contents requires resealing to the original standards, which is usually difficult to ensure.

BARRIER MATERIALS

Waterproof and water-vapour-proof barriers may consist of an inner film wrap of polyethylene, polyvinyl chloride, waxed or bitumenised paper, metal foil, or composites of foil, plastic and paper, but it may also be provided by the main structure of an enclosure particularly if this is of metal or rigid plastic. The method of sealing and its efficiency is important. Most of the plastic and composite films can be welded, and the proofed paper products sealed, to give reliable closures; resealable closures are available and have attractions but are sometimes a source of problems.

Plastic barrier films are widely used to reduce the transmission of water vapour into the interiors of packages; with desiccants they give a high standard of protection to metals. Three classifications of barrier materials are recognised:

(1) waterproof: highly resistant to the passage of liquid water;
(2) water-vapour-resistant: WVTR at 38°C from 90% RH to 2% RH is no greater than 8 $gm^{-2}d^{-1}$ per day;
(3) water-vapour-proof: WVTR at 38°C from 90% RH to 2% RH is no greater than 1 $gm^{-2}d^{-1}$ per day.

WVTRs are measured under two conditions for packaging, at 25°C from 75% RH to 2% RH representing temperate conditions, and at 38°C from 90% RH to 2%RH for tropical conditions. WVTRs of some typical samples of the most frequently used barrier materials are given in Table 14.1 The

Table 14.1 — Water vapour transmission rates (25 μm film)

Material	WVTR ($gm^{-2}d^{-1}$)	
	Temperate 25°C and 75% RH (into 2% RH)	Tropical 38°C and 90% RH (into 2% RH)
Metal foil (new)	0	0
Vinylidene chloride copolymer	0.1	1
High-density polyethylene	1	9
Polypropylene	4	12
Low-density polyethylene	5	18
Rubber hydrochloride	5	15
Polyester (linear)	10	25
Rigid PVC	10	25
Flexible PVC	25	100
Nylon 6	59	200
Polystyrene	70	180
Cellulose acetate	500	1000

WVTR is approximately inversely proportional to thickness, so that results for other thicknesses can be simply derived from Table 14.1.

Perfect metal foil, or metal foil/plastic laminate, should not transmit water vapour, and when new carefully prepared samples are tested, their WVTRs are very low, but the values increase very rapidly if the films are folded, creased or flexed, so that these materials are only satisfactory as impervious liners on inflexible non-reusable containers. When a sealed flexible liner is required, plasticised PVC is most frequently selected, although polythene, polyester, polyvinylidene chloride and rubber hydrochloride are also used. Although the WVTR of PVC is initially higher than these alternatives, it is less liable to increase as a result of flexing, ageing or contamination with oils or solvents. PVC can also be reinforced by fibres and joined by high-frequency welding; a good account of these and other properties of PVC, including a reusable closure which has a WVTR of 0.17 $gm^{-2}d^{-1}$ at 38°C and 90% RH is given by Bayley (1970).

DESICCANTS

Desiccants are materials which can absorb a high proportion of water either chemically, or physically within the pores of an open structure; the latter type are the most widely used. Desiccants for use in packages must be inexpensive, easy to handle, non-corrosive and efficient in terms of the quantity of water they absorb; those which best meet this requirement are silica gel, activated alumina and bentonite clays; where very low humidity is needed the extra cost of ion exchange resins (molecular sieves) may be justified. Calcium oxide (lime) or a mixture of calcium and sodium oxides (soda lime) is used for reducing the humidity within store rooms. Desiccants typically absorb 25% of their weight of water but this varies with the type over the range 15–35%. Their capacity is rated againt a standard silica gel which has an absorptive capacity of 27% of its weight as water in an atmosphere of 50% RH.

Silica gel is the most widely used desiccant; it absorbs up to 30% of its weight of water, and is available as beads or granules. Absorbed water can be removed, and the desiccant regenerated, by heating above 140°C. Cobalt chloride is often added to silica gel as an indicator of the extent of water pick-up; the gel is blue in equilibrium with air of RH below about 30%, and pink in equilibrium with air of RH above about 70%, but the colour is too imprecise for use as more than a general indication.

Activated alumina absorbs about 16% of its weight of water at 50% RH. It can be reactivated by heating at 250°C. It is used in a similar way to silica gel.

A molecular sieve, a synthetic zeolite, of a uniform pore size of 4–5Å, absorbs up to 15% of its weight of water at 10% RH. It can be reactivated by purging with dry gas at 250°C. It is relatively expensive and its use is restricted to specialist applications, mostly critical electronic assemblies, which need to be kept very dry.

It is important to monitor the air in a desiccated container to ensure that the desiccant remains effective and for this purpose indicator papers based on cobalt salts are often used. Where it is necessary to monitor the humidity more closely, hair hygrometers or wet and dry bulb thermometers have been used but are inconvenient. More recently electronic systems based on changes of conductivity of salt-impregnated bridges or the dielectric properties of certain plastics have been developed as continuous, or plug-in, monitoring devices.

Desiccants are available as granules or beads and should be contained within the package in dustproof bags. They are best positioned within perforated metal containers which should be secured to the equipment or its supports. It is sometimes advantageous to secure the desiccant to the outer container with resealable access from outside so that the desiccant may be replaced.

The quantity of desiccant needed in a package is often derived from the semi-empirical relationship:

$$W = 70ARM + 0.5D$$

where: W is the weight of desiccant needed in grams (based on 27% absorption of water)
A is the total area of the barrier in square meters
R is the WVTR of the barrier material (using the temperate or tropical measurement as appropriate)
M is the required life in months
D is the mass of hygroscopic material in the package

For very large containers the cost of desiccants may be great and when the area of the barrier film exceeds about 100 m^2 it may become more economical to maintain the dry atmosphere by air conditioning. Since each 11°C reduction halves the saturated vapour pressure of air, cooling air at 50% RH by 22°C removes about half its water content, if equilibrium conditions are attained, and allows a ready control of humidity. A temperature reduction of 20°C is a practical minimum for economic operation. Cooling may need to be coupled to some heating to give an acceptable environment. Air conditioning is preferred in large storage areas particularly in moist tropical, or semi-tropical, conditions.

Although desiccation prevents most corrosion, there are exceptions, notably the sulphide tarnishing of silver and copper, fretting corrosion, the growth of metallic whiskers, attack of aluminium by mercury, the allotropic transformation of tin at low temperatures, and failure of steels from hydrogen embrittlement under stress. It should also be remembered that although corrosion does not proceed at low humidities even if contaminants migrate to the metal surface, once the contents encounter a humid environment the effects of contamination may become apparent as rapid corrosion ensues. Electrical contacts plated with precious metal which become contaminated by organic condensates during periods of desiccated storage sometimes suffer an unacceptably high increase in contact resistance as a result of the film of impurities which builds up on the surface.

REFERENCES

Bayley, F. H. (1970) *Proc. Conf. Protection of Metal in Storage and Transit,* London, Brintex Exhibitions, p. 73.

Middlehurst, J. & Kefford, J. F. *Proc. Conf. Protection of Metal in Storage and Transit,* London, Brintex Exhibitions, p. 83.

Evans, U. R. & Taylor, C. A. J. (1974) *Br. Corros. J.* **9,** 26.

15

Packaging materials

The packaging industry as a consumer of materials is enormous, varied and growing fast. In the UK alone about £4000M was spent on packaging materials in 1984. Food and household products accounted for 80% of this total, while about half of the remainder (i.e. £400M) was used in packaging materials for general engineering products. Ensuring long life and durability of materials used in packaging is a constant challenge to materials technologists but the problems of copious litter and high consumption of raw materials are of increasing concern, and emphasis is being directed to recycling materials or ensuring that they will degrade either chemically or biologically on disposal.

Packaging materials may be grouped broadly into four main categories: wood products, plastics, metals and glass. The expenditure on each of these in 1984 was respectively £1400M, £970M, £960M and £440M with miscellaneous items amounting to about £120M. The quantities of the first two categories used in packaging are steadily increasing. The total value of metal products used is static with a swing towards foil and thinner gauge strip steel and aluminium, while that of glass is falling. Natural products such as wood, wool, cotton, jute, hair, straw and adhesives based on natural materials are steadily being supplanted by synthetic polymers, although processed wood products more thanmaintain their market share.

WOOD PRODUCTS

Unprocessed timber now represents no more than about 6% of the total value of wood products used in packaging, the remainder consisting of corrugated and solid fibreboard, cartons, flexible wraps, paper sacks and bags, moulded pulp containers and loose-fill insulation. Wood itself is best avoided in packaging metals because of the risk of vapour corrosion (see Chapter 6). If it has to be used either as an outer container, or as internal furniture, effective barriers and protectives are needed.

The physical properties of wood and wood products depend on the cellulose fibres which have good longitudinal strength and readily bind together to give cohesion between fibres. The low cost of the raw material, the ease of processing, and the desirable and versatile properties of the final product all contribute to the strong position of wood products in packaging. The ease with which paper products can be disposed of after use by recycling or by biodegradation is a factor making them find increasing favour. Although plastics are displacing wood products from some traditional uses this is probably more than compensated by the new and desirable combination of properties available from new uses of wood or paper as laminates and composites.

Wood may be broken down and pulped mechanically to its constituent fibres and then formed into various qualities of board, or it may be treated chemically by the sulphite and sulphate process to give pure cellulose fibres. The lignin dissolves as sulphates and sulphonates, and the acetyl and methyl side groups hydrolyse and dissolve in the sulphite liquor. The pulp is washed, layered and consolidated. The mechanically pulped board is inexpensive while the chemical pulp gives a bleached white product relatively free from soluble impurities. Composites of white pulp on a substantial substrate of board give a popular compromise of an attractive surface which accepts a high standard of printing and decoration, with bulk and strength provided by the inexpensive mechanical pulp. Layers of plastic and/or metal foil may be laminated with paper or board, or resin may be incorporated into the final pulp, to give improved stiffness, strength, electrical insulation, water and oil resistance, or resistance to diffusion of water vapour or odours.

In packaging of metals, wood itself has the disadvantage, discussed in Chapter 6, that it is a potent source of corrosive vapours. Hot treatments to force-dry wood either initially from the green state, or after aqueous impregnation with rot- or fire-proofing salts, or hot bonding or steam bending, give further quantities of acetic acid and render the wood more corrosive.

Chemically pulped paper is free of acetic acid but sulphate and chloride residues from the processing chemicals may cause corrosion of metals in contact with the paper (see Chapter 6). If sulphate and chloride residues are reduced below 0.25% and 0.05% respectively (calculated as the sodium salts) the paper is much less liable to cause corrosion and may be used as a primary wrap for metals. Positive protection may be afforded to metals if the paper is impregnated with inhibitors; volatile inhibitors give protection to some distance from the paper while non-volatile inhibitors protect only the surface in contact (see Chapter 13). Papers and boards lose much of their strength when wet and are liable to rot at high relative humidities; they also need protection against attack by insects and rodents.

Wicker baskets still find wide use as outer protectives, especially for glassware and as containers. They are especially light, resistant to impact and versatile although they do not readily lend themselves to mechanical handling and stacking.

PLASTICS

Plastics are grouped in two major classes, thermoplastics and thermosetting, or cross-linked, polymers. Both types have very large molecules with molecular weights of up to several million.

The thermoplastics contain mainly linear molecular chains and therefore soften on heating so that they can readily be formed by injection moulding, extrusion or rolling into a variety of shapes. The largest class of thermoplastics is formed by free radical polymerisation of vinyl monomers. The main members of this class are listed in Table 15.1.

Table 15.1 — Some vinyl plastics (— $C(R_1)_2$–CR_2R_3 —)

R_1	R_2	R_3	
H	H	H	Polythene
H	H	CH_3	Polypropylene
H	H	$O.CO.CH_3$	Polyvinyl acetate
H	H	OH	Polyvinyl alcohol
H	H	C_6H_5	Polystyrene
H	H	Cl	Polyvinylchloride (PVC)
H	Cl	Cl	Polyvinylidene chloride
F	F	F	Polytetrafluoroethylene (PTFE)
H	CH_3	CO_2CH_3	Polymethymethacrylate (Perspex)

Mixtures of monomers can be allowed to react together to give copolymers, so that for example vinyl acetate and vinyl chloride together provide a softer plastic than PVC but with otherwise similar properties. Thermoplastics are sometimes softened by incorporating non-volatile organic liquids as plasticisers. Thus PVC plasticised with di-octyl phthalate gives a soft plastic often used as a thin wrap in flexible containers. Thermoplastics may also be mixed together to obtain improved properties, as has been done so successfully with styrene–acrylonitrile copolymer and polybutadiene to give the tough impact-resistant, engineering plastic ABS (acrylonitrile–butadiene–styrene), which also has the unique property among commerical plastics of accepting, after appropriate processing, an adherent metal electrodeposit (usually nickel/chromium).

Important thermoplastics are also made by condensation reactions: most importantly, diamines with dicarboxylic acids giving polyamides (nylons), glycols with saturated dicarboxylic acids giving linear polyesters, aromatic diols with carbonyl chloride to give polycarbonates.

The thermosetting polymers are formed by cross-linking reactions, between a monomer having two reactive locations, and short-chain intermediate polymers with reactive positions along their length. The net effect is to give a cross-linked rigid network which, once formed, cannot be softened, so that the reaction mixture has to be cast before reaction is complete. Important thermosetting polymers include the polyesters, epoxides, polyurethanes, phenolics and formaldehyde resins. They are inert and hard but inclined to be brittle and their properties are often improved by incorporating fibre, paper and fabric reinforcements and by powder fillers.

Plastics are used in packaging as barrier films, containers, structural supports and cushioning. They are relatively inexpensive, strong and flexible, and can be readily produced in a wide range of sizes and dimensions in attractive colours or with near 'water white' transparency. They can be resistant to oils, water, chemicals and biodegradation. Their desirable properties, versatility and low cost, have produced a revolution in packaging and presentation. However their limitations should not be forgotten. They are permeable, soft, change their properties with age, lack dimensional stability, have poor resistance to heat, ultra-violet radiation and many organic solvents, and are difficult to specify uniquely.

Polythene is the plastic used in greatest quantity in packaging followed by PVC and thermoplastic polyesters. Other plastics widely used in packaging, although in lesser quantities, include ABS, polypropylene, polystyrene, polyvinylidene chloride, nylon, unsaturated polyester and epoxides reinforced with glass fibre, polyurethanes, chloroprene rubbers, formaldehyde resins, cellulose acetate and ethyl cellulose. Each generic formula includes a wide variety of individual formulations and processing procedures to give properties suited for specific environments and applications.

The provision of water-vapour-proof barriers is arguably the most important use of plastics in the packaging of metals. For this use, water vapour transmission rates (WVTR), given in Table 14.1, are vital criteria.

Other important criteria are strength, flexibility, ease of sealing, chemical stability, resistance to biodegradation, compatibility with other materials and cost. One of the problem plastics in the protection of metals is PVC. Its virtues obviously outweigh its vices since it is second only to polythene in tonnage used, but it is the least stable of the synthetic polymers and when the basic polymer does degrade the resultant volatile, hydrogen chloride, is very corrosive to many metals. Chemical degradation of the polymer chain (unzipping) is combatted by chemical stabilizers which are generally effective until the temperature rises to about 60°C when degradation starts, becoming marked at 70°C, and once begun accelerates quickly since the reaction is autocatalytic. Degradation is accelerated by sunlight and high humidity. It is accompanied by colour changes, first to yellow, steadily deepening through orange to brown. The ester plasticiser (often di-octyl phthalate), widely used to provide flexibility to PVC, is less expensive and more effective than alternatives at lower temperatures but has the disadvantage of readily migrating into other materials and sometimes reacting with them. Contact of PVC with painted surfaces, or zinc or copper alloys, is particularly to be avoided since the paint may be softened, zinc may catalyse depolymerisation and copper may react with ester plasticiser to give bright green corrosion products; the corrosion product in turn may assist in degrading the PVC polymer. Nitrocellulose and PVA are two other plastics which degrade chemically to give corrosive vapours. The nitric acid decomposition products from nitrocellulose are very corrosive to a wide range of metals and the instability of this polymer, coupled with its high inflammability, make it an unattractive option for any but a few specialised applications.

Some plastics contain mobile residues after manufacture, which may later migrate to cause corrosion. Cross-linked polyesters, PVA, formaldehyde resins and polyurethane give traces of volatile organic acids and aldehydes. These volatiles not only cause corrosion and form troublesome films on metals and semi-conductors but are also receiving increasing attention, as a cause of various distressing symptoms from allergic reactions of sensitised individuals, tainting foodstuffs and adversely affecting the characteristics of explosives and other reactive compostions. The volatiles evolved from plastics in fires are a related phenomenon that has attracted considerable interest since the hydrogen chloride and other volatile halogens and cyanides evolved are usually a greater risk to life than the fire. Many of these degradation products are also very corrosive, and secondary effects from subsequent corrosion are often more damaging than the initial damage from small fires or from components burnt out through overloading in electrical and electronic assemblies.

METALS

Most of the metal used in packaging is in the form of food cans and foil, but it is also important for drums and in the construction of other robust external containers which can be sealed to high standards to give the best protection to sensitive or valuable components. Steel and aluminium are the constructional metals most used, with zinc, tin and chromium metals often applied to the steel as protective coatings.

The construction of cans for food products is an evenly matched contest between tinned and tin-free steel with a marginal use of aluminium. Tin-free steel is protected by coatings of chromium metal, chromium oxide and oil. Improvements in forming, joining and protection are constantly being exploited to swing the economic advantage first one way, and then the other. The competition for sealed drink containers is even wider, with steel, aluminium, glass, plastics, and paper composites all involved in offering attractive properties and lower costs, to retain or increase market shares. Tinned steel is generally preferred for containers of fruit products for which it is peculiarly well suited, since not only has tin virtues in easy forming, soldering, welding and offering a good basis for protective lacquers, but it also provides a protective coating which is low in porosity and is attacked only slowly by fruit juices and yet provides sacrificial protection to the base steel at pores, scratches and other imperfections. The sacrificial protection given by tin in this environment is in marked contrast to its performance on steel in atmospheric exposure when it is cathodic to steel and gives no protection at pores; this lower potential of tin in fruit juices arises through the tendency of tin to form complexes with the organic acids present so that the activity of the tin cation in solution remains low.

Aluminium metal foil is a strong growth area among packaging materials. It provides an impervious wrap which does not stain, taint or react with foodstuffs and is inert in many environments. It finds use in conjunction with plastics and paper as a laminate which combines the flexibility and robustness of paper with the water resistance of plastics and the barrier properties of metal. Metal foil is sometimes also used as a final layer in plywood to prevent water vapour and acid vapours from the outer layers of wood penetrating into the package. Although metal film is impervious when it is manufactured, pin holes do develop at folds, wrinkles and sites of wear or corrosion, so that measured vapour transmission rates into assembled packages are often disappointingly high.

GLASS

Glass is a supercooled liquid consisting of silica with sodium and potassium silicates, which has so very many attractive properties that it is at first surprising that its use in packaging is steadily falling. The explanation lies in its two overriding vices. It is heavy and brittle. Its desirable properties are however worth recalling. It is inert, impermeable, reusable, easily sterilized, made from cheap raw materials, rigid and strong. It gives no taste; it emits no toxic, corrosive or otherwise noxious substances; it resists all solvents (apart from acid fluorides); it can be transparent, coloured or opaque; it can be used as a measuring container; it can be obtained in a variety of shapes; it does not discolour, and it has very good dimensional stability.

The tensile strength of glass without surface imperfections is 7000 MN/m^2, but as normally produced its practical tensile strength is 30–70 MN/m^2. Stress concentrates at the tip of microcracks on the surface, and once these cracks begin to propagate there is no mechanism equivalent to the yielding which blunts growing cracks in metals and plastics. Two methods are available however for exploiting, at least in part, the high tensile strength of glass: the use of glass fibre reinforcement in plastic where the glass surfaces are protected, and cracks only propagate through a single fibre; and arranging that the surface is maintained in compression. Surface compression can be introduced by rapidly chilling the surface of hot glass, by surface treatment of the cooling glass with tin compounds and then organic lubricants, or by treating the surface with potassium phosphates so that a proportion of the sodium ions are replaced by potassium. These 'toughening' treatments require an appreciable thickness of glass and this limits the extent to which toughening can be coupled with weight reductions. Plastic coatings and sleeving help to lessen the risk of breakages by protecting the surface and by reducing the peak loading from impacts.

STORAGE OF PACKAGING MATERIALS

Close control of storage conditions for all new materials to be used in packaging is essential. They should be stored separately from packaged goods or returned empty packages and should be segregated according to composition and type. A first requirement is an enclosed building free of rodents and with as little harbourage as possible for insects and moulds according to the principles outlined in Chapter 9. Regular fumigation against insect pests may sometimes be advisable but advice should be taken on the compatibility of the chemicals used with those being treated or likely to come into contact. Air circulation should be maintained with filtered intakes, away from any furnace outlets or other sources of contaminants. The whole area should be kept clean and sources of dirt and debris should be removed or repaired; possible entry points for rodents, insects and birds should be covered with fine wire mesh.

Adequate stocks of all materials will need to be maintained but care is needed to ensure that materials are not held over long in store; withdrawals should be from oldest stock. Access should be limited to authorised personnel and records of stock removal maintained. Precautions may be needed to prevent unauthorised withdrawals.

The temperature within the storage area should be maintained within the range of 15–21°C if possible, and even more importantly the relative humidity kept within the limits of 45–60%. Storage of materials in direct sunlight or near a source of radiant heat is to be avoided.

Materials should be stacked clear of the ground on shelves, racks or pallets. Empty containers should be stored upside down with lids off, or on their sides with lids on. Rolls of wrapping materials should be stored on end in their original wrappers with positive restraint against the risk of toppling. Rolls of gummed and self-adhesive tape must be stored on their sides. Desiccants should be stored in air-tight containers. Temporary protectives, especially those containing solvents, should be stored in closed containers. Papers impregnated with volatile corrosion inhibitors (VCIs) should be kept in sealed wrappers.

In stacking containers due regard should be given to loading at the base of the stack; thermoplastics in particular may assume a permanent set on sustained loading with relatively low loads, and permanent creases form at folds in wraps on long storage. Space must be provided around stacked materials to allow access for maintenance, inspection and withdrawal, and, if available, for mechanical handling equipment.

16

Packaging for Corrosion Protection

DESIGN FOR CORROSION PROTECTION

Corrosion prevention starts on the drawing board, as an integral aspect of design, from the choice of materials, through methods of manufacture and finally to ensuring that the chosen system of metal preservation is on a sound basis — metal surfaces that have been adequately cleaned, prepared and handled in a way that has left them free of active contaminants, and then pretreated and appropriate protective finishes applied. Rules to be followed in meeting these ends are:

- use specified materials and processes as far as possible, ensuring appropriate thicknesses and precautions are detailed (see Annex for a selection of guides and specifications related to the protection of metals in storage). If unspecified or proprietary materials are used, some testing and evaluation may be needed and a process specification incorporating quality control testing should be drawn up and followed. Treat suppliers' unsubstantiated assurances of good performance with caution;
- design to meet the harshest aspects of the user environment but also for the environments of transport and storage;
- be clear on the extent to which corrosion protection is to be built in, or will depend on maintenance, conditioned environments or temporary protection;
- avoid crevices where possible, and where they occur at joints, threads, inserts and flanges apply chromate-based jointing compositions;

— radius corners, edges and re-entries. Remove machining burrs, and smooth surfaces;
— do not stamp identifications onto finished metal parts;
— leave sufficient tolerance for protective finishes;
— ensure that assembly procedures include protective treatments in appropriate sequences;
— avoid undesirable bimetallic contacts. Insulating inserts are a possible alternative (note that graphite, carbon-reinforced plastic, for example, acts as a noble metal in contacts);
— provide adequate access into enclosures, and drainage from their lowest points (test properties of finishes on the internal surfaces themselves, not on flat test-pieces processed at the same time). Sealing all entries into enclosures is an alternative in theory, which is difficult to achieve in practice;
— avoid, if possible, alloys subject to stres corrosion. If they are used, select a heat treatment that gives least risk, heat treat to relieve any strains introduced during fabrication and avoid force fits or distortions in assembly. Sacrificial metal coatings provide the best protection against stress cracking;
— allow for de-embrittling baking procedures if high strength steels are processed in aqueous solutions or by other treatments liable to result in pick-up of hydrogen (see Chapter 5). For the application of metal coatings to high strength steel, vacuum deposition is to be preferred to electrodeposition;
— springs may present severe problems in storage and care is needed in selecting materials and protective finishes (see for guidance DEF STAN 01-4/1 and BS 1726 on the selection of metallic materials for springs, and DG-10 on their protection);
— if possible use hermetic sealing or encapsulation for electronic circuits;
— be alert to the risks presented by hygroscopic materials, and materials which may act as wicks and wet compacts;
— consider the risks from possible incompatibility, especially from organic materials with magnesium, steel, zinc, cadmium and lead; chloride and sulphate in wraps with metals; plasticised PVC in contact with paint and other organic materials, and long-chain fatty acids, esters and sulphur-containing materials with copper alloys;
— ensure the processes, materials and products are adequately monitored and the results recorded.

APPROACH TO PACKAGING

The purpose of any package is to contain, and to provide protection and information. These functions interact and their relative importance varies but failure of any one is likely to be serious and possibly disastrous. A first consideration in designing the protective aspects of a package is to analyse the risk from all possible hazards and set them in an order of priority. The overall design of the package is then directed, within the constraints of the other functions required of it, i.e. cost, and the handling assets and limitations available during transport, to protecting the contents from the primary hazard. Other hazards are then combatted on a basis of descending priority according to the levels of risk associated with them. Corrosion is countered within this framework.

Corrosion hazards have been reviewed in Chapter 2. Time, temperature, relative humidity, temperature variations and the amounts and type of active contaminants are the major factors which determine the extent of corrosion. The degree to which corrosion can be tolerated, the consquence if it does occur and the susceptibility of the design to corrosion, are the major factors in setting the priority of corrosion protection.

Protection from corrosion should be provided as close as possible to the metal surface. Assessment of the risk is linked to a knowledge of the susceptibility to corrosion of the metals involved, and the effect of corrosion on functioning and appearance. Rust on robust items, such as crowbars and pickaxes for instance, is unlikely to impair their functions, even after prolonged storage in bad conditions, although a customer may find their appearance unacceptable, but a heavy patina would probably be accepted without comment. This same level of corrosion would ruin more critical items such as measuring guages, bearings or clocks. Many hand tools tarnished to the same degree would function but their appearance would be unacceptable. Much lower levels of tarnish would give failures in the more delicate mechanisms of clocks, watches, cameras and precision bearings, or would be unacceptable on surgical instruments or decorative ware. Even an invisible thickening of the surface film on electronic items may affect reliability and performance. It should also be remembered that even if contamination during storage does not lead to obvious corrosion before delivery, it may cause corrosion later or interfere with any further finishing processes.

PACKAGING TO PROTECT AGAINST CORROSION

The standards of packaging, in increasing levels of effectiveness in the protection they provide against corrosion in storage and transit, are:

(a) no additional protection — either the permanent protective is relied on for protection or corrosion is tolerated;
(b) temporary protectives only: sometimes with wraps applied to more critical areas, with or without slatted crates or pallets, or fibreboard containers to provide mechanical protection and allow for ease of handling;
(c) a single primary wrap with temporary protectives, with an external container;
(d) as (c) but with sealed water vapour-proof barrier;

(e) water vapour-resistant barrier with desiccation;
(f) hermetically sealed in atmospheres of dry air or dry nitrogen.

The reader will now have gathered from the description of the hazards, failure modes and protective treatments in the book so far, the general philosophy of preventing corrosion in packaging. But at the risk of some repetition the essentials of the approach are summarised in the remainder of this section.

Wood presents a high hazard in packages and enclosed storage areas for steel, zinc, cadmium, magnesium or lead. If wood is to be used in the packaging of metals, oak or sweet chestnut must be avoided and some testing is advisable to eliminate timber of high acid content. Precautions should also be taken to prevent corrosive vapours from any wood present reaching metal surfaces — by slatted construction of wooden outer containers for instance, which allows ventilation, or by barriers between the wood and stored metal. Aluminium foil on the inner surfaces of plywood, and inner wraps of PVC, polythene or bitumenised paper all assist in reducing the hazard to a low level. Other potent sources of corrosive vapours are inadequately cured phenolic/formaldehyde resins, cross-linked polyesters and drying-oil based paints. Because of the hazard of vapour corrosion the fittings and internal surfaces of packages for metals should not be painted with drying-oil paints. Although more expensive, epoxide and polyurethane paints — of the two-pack, or the catalytically cured stoving varieties, not the drying-oil modified variants — give a higher standard of protection than drying-oil paints and do not give corrosive vapours. Their use should be considered on higher value items which are likely to remain enclosed for long periods.

Any materials in contact with metals during storage should be inert and free of corrosive impurities (see Chapter 7). Attention should be given to possible compatibility problems from paper and plastic wraps, plastic and wooden supports, and from fillers and cushioning materials.

Having established an environment low in corrosive contaminants on and around the metals to be protected, and assuming that the physical hazards have been combatted, then the external hazards which may cause or intensify corrosion are: high relative humidities, high temperature, wide variations of temperature, sunshine, rain, salt spray, corrosive atmospheres, soot and other debris, biodeterioration, and attack by rodents and insects. Microbiological attack is controlled by biocides, by good housekeeping and the selection of resistant materials as discussed in Chapter 9, and by avoiding warm damp conditions. Protection from rain, sun and most of the salt spray and soot should be provided by external shelter — a storage building, container, the body of a vehicle or ship, or at least a tarpaulin cover. A package, even one made from relatively flimsy materials and poorly sealed, should exclude sunshine, liquid water and solid contaminants under even primitive cover. The remaining hazards are from acid vapours in the atmosphere, high temperatures, high relative humidities and temperature cycling.

The total hazard is a function of exposure time although not a linear one. Once deterioration has begun it accelerates and its effects often spread rapidly.

The types and standards of wraps and barriers do little to reduce the risks from fretting, or brinelling, which is unaffected by humidity or contaminants and has to be considered as a separate risk from other types of corrosion. Bearings, sliding joints, leaf springs and surfaces of stacked metal boxes or metal sheets, and the strands of wire ropes of winding gear are all susceptible to fretting. The risks to bearings may be reduced by cushioning or by some support of the upper part of the assembly. Fretting between the surfaces of stacked metal sheets is frequent and damaging. The impression of an inclusion near the base of a stack can be transmitted by fretting through the whole stack. A basic precaution is to ensure that sheets are stacked on flat surfaces free of marked imperfections or debris, and that no inclusions are trapped between them. To prevent fretting, surfaces should be treated with a temporary protective or separated by paper (papers must be low in impurities, and soaps, organic acids and esters must be avoided near copper or brass). When the expense of treating surfaces is unacceptable, such as with bulk quantities of tinned steel sheet, and tight binding with wedging is used to prevent the relative movement, there is a risk of exacerbating the effect by reducing the amplitude of the movement without stopping it entirely, while at the same time increasing the loading. Even when experience has established a degree of binding which prevents fretting, consideration must be given to any changes in transport, in the size or weight of the stack or of containers, since relatively small changes can move the system into a resonating state where damage increases rapidly.

High strength alloys, especially those susceptible to stress corrosion or hydrogen embrittlement, need careful consideration (see Chapter 5 for methods of reducing susceptibility to stress corrosion and hydrogen embrittlement). Springs present particular problems since they are of necessity made from strong alloys and they may sometimes have to be stored in a stressed state, and often because of their small dimensions or close tolerances they are only lightly protected.

ENVIRONMENTAL TESTING OF METALS AND PACKAGES

The likely performance of packaged goods can be determined by prior environmental testing. Four classes of risks may be distinguished and investigated by appropriate testing:

— corrosive atmospheric environments;
— impurities already present on the metal surface;
— incompatibility between metals and nearby materials;
— corrosion as a secondary consequence of the deterioration of other materials;

ATMOSPHERIC CORROSION

Numerous protective treatments are available for metals. Their full evaluation requires extensive exposure trials, but rapid and simple tests are available which reveal weaknesses and give some measure of corrosion resistance. Two typical and particularly simple tests for resistance to atmospheric corrosion which have been widely used as quick confirmatory tests for temporary protectives and passivation treatments used to protect goods in storage are described in BS 1391 (1952): the salt droplet test and a sulphur dioxide test, the CRL beaker test. The apparatus for these two tests is shown in Fig. 16.1 (a) and (b).

The first of these, the salt droplet test, simulates contamination in marine atmospheres and the second, sulphur dioxide contaminated urban atmospheres. Although BS 1391 has been withdrawn, these and other simple tests are still used in some specifications and results from the salt droplet test have been quoted in several places throughout the book. They are being replaced by internationally agreed tests. Some tests have been developed specifically to correlate with certain types of user environments. The CASS and Corrodkote tests, for instance, were developed to give the types of failure encountered in service on chromium/nickel-coated steel components on motor cars. Established tests of this type are described in BS 3745, 5411 and 5466. A very good account of corrosion testing and correlation with experience in use has been given by Shreir and LaQue (1976).

With high strength alloys of copper, aluminium, titanium and iron (especially stainless steels), stress corrosion is a hazard and a range of exposure tests with bent beams or slow tensile tests, sometimes on pre-cracked specimens, have been developed. There is however a fundamental difficulty in devising a general test for a phenomenon which is usually restricted to very specific, and sometimes unusual, environmental conditions. Fortunately experience and the present understanding of the characteristics of stress corrosion usually allows suitable tests to be devised. With certain alloys specific tests have been developed which reveal the metallurgical state likely to give stress corrosion. Typical tests of this type include the mercurous nitrate tests for brass and the Moneypenny Strauss test (see BS 5903) for austenitic stainless steels.

Generally the results of any accelerated test programme should be followed by and correlated with exposure in outdoor environments. For packaged metals, covered exposure tests in marine or urban environments are usually appropriate.

CORROSIVE RESIDUES

The presence of corrosive contaminants on metal surfaces from the variety of causes described in Chapter 8 is revealed by exposure in an uncontaminated humid atmosphere — 30 days at 30°C and 85% RH, which are typical conditions for developing corrosion or staining on surfaces contaminated with aggressive residues.

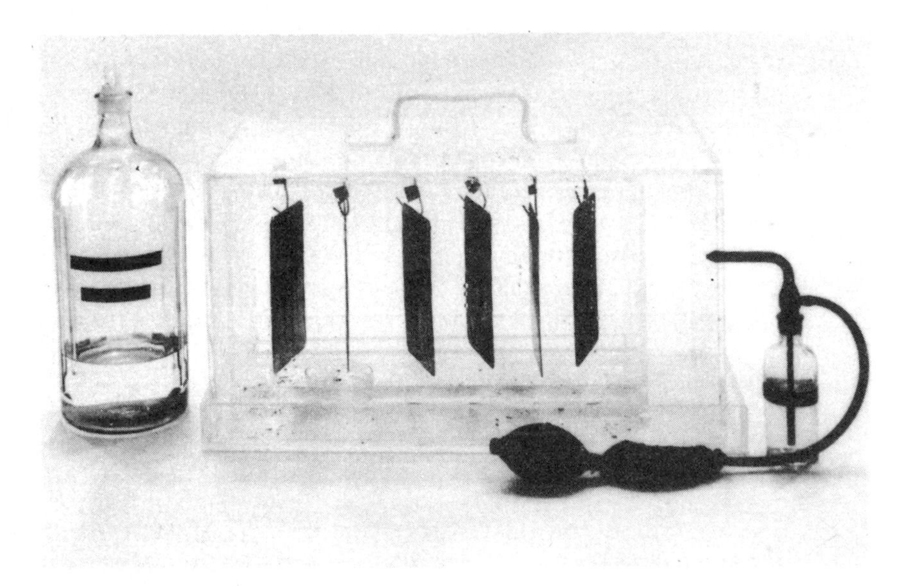

Fig. 16.1 — Corrosion test apparatus of BS 1391 (1952). (a) SO_2 beaker test; (b) salt droplet test assembly. (© Controller, Her Majesty's Stationery Office 1986).

ENVIRONMENTAL TESTING

Other corrosion hazards usually become evident during environmental testing of packaged items, although specific risks such as the effects of wraps on metals, or vapour corrosion by particular plastics may be evaluated in subsidiary tests on metal coupons in humid atmospheres. Corrosion by incompatible materials is revealed by exposure to warm damp atmospheres such as the conditions of BS 2011 (Part 2.1Ca), in which the environment is maintained at 40°C and at 90–95% RH for a period of 4, 10, 21 or 56 days. A more searching test, especially for revealing vapour corrosion effects, is provided by BS 2011 (Part 2.1Db), in which the temperature is cycled between 25°C and either 40°C or 55°C with the humidity at 90–95% RH. The cycle is 24 hours with the upper temperature being maintained for 12 hours, with 3 hours allowed for each temperature change. The higher maximum temperature of 55°C represents the expected upper storage temperature of tropical exposure and the 40°C maximum is for temperate conditions. Maximum temperatures in environmental tests for defence equipments are 5–10°C higher and these tests are therefore usually more exacting.

Secondary corrosion sometimes occurs in mould growth testing (BS 2011: Part 2.2J) but is more generally encountered during sequential testing in which damp heat follows mechanical stressing or heating, in which seals, wraps or surface coatings may have been damaged, or organic materials partly decomposed.

Fretting corrosion may occur in vibration tests. Other types of surface deterioration such as the growth of metallic whiskers are not easily induced by environmental tests, but careful inspection after a relatively short period may give an early indication.

The effects of corrosion are sometimes revealed in function tests after environmental testing or will be seen on inspection. Careful visual inspection after a relatively short period of testing to identify the processes of breakdown is often more informative than the more obvious gross effects which develop after prolonged exposure. Inspection is best carried out by a skilled and knowledgeable observer looking for such effects as incipient stress cracking, the first sign of fretting, patterns, blisters and breakdown of coatings, staining from residues, dezincification of brass, corrosion in cavities, or the first white dusty appearance of vapour corrosion attack on cadmium or zinc, all of which are readily overlooked by the untutored eye.

QUALITY ASSURANCE

Little specific reference has been made throughout the book to quality assurance, since quality is nowadays rightly seen as being assured only through overall competence in, and by efficient integration of, all aspects of design, production and distribution. The manifestations of quality in engineering products are in their appearance, durability, reliability, availability (i.e. available for use and not undergoing servicing or repair) and maintainability. In each of these aspects corrosion and its prevention is important and the total theme of the book is best seen as a contribution to quality assurance.

If corrosion failures do occur in production, storage or use, it is essential that methods of analysis, diagnosis and evaluation are readily available, so that corrective action may be devised, evaluated and fed back into the design and production system. It is equally vital that the occurrence of rejects or failures is prevented by appropriate and quantified monitoring at all stages from the start of production to final delivery, and where possible, by monitoring throughout subsequent use. Any trends away from the design optimum can then be identified from the monitoring test results before rejects appear, and corrective action can be taken. To support this approach complete systems of recording and tracing materials from receipt and through all stages of processing and assembly, together with detailed procedures for all processing, should be instituted. These procedures provide a structure for building up and maintaining a standard of quality, and for tracing the implications of any variation through the system. This process of feedback and learning provides a positive approach to quality that allows continuing improvements.

Experience shows that most failures result from either ignorance or a lack of control at interfaces between areas of responsibility — everybody's business can rapidly become nobody's business. Metal finishing, and protection in storage and transit are areas which suffer from both of these causes since the scientific understanding is not widely distributed and both the technologies and processes involved come at critical interfaces. It is hoped that this book will disseminate the information to those involved in these areas and assist in obtaining improvements in the quality of goods, and in 'value engineering' their design and production.

REFERENCES

Shreir, L. L. & LaQue, F. L. (1976) Shreir, L. L. (Ed.) *Corrosion*, Vol. 2. Newnes-Butterworths.

Annex: Specification and Guides on Packaging, Environmental Testing, Protective Processes and Coatings

BRITISH STANDARDS

(Obtainable from British Standards Institution, Linford Works, Milton Keynes M14 6LE.)

BS 903 Part A33 (1977) *Methods of test for staining in contact with organic materials.*
BS 903 Part A37 (1979) *Assessment of adhesion to and corrosion of metals.*
BS 913 (1973) *Wood preservation by means of pressure creosoting.*
BS 1133 *Packaging Code.*
Sections 1–3 (1967) *Introduction to packaging.*
Section 4 (1965) *Mechanical aids in package handling.*
Section 5 (1964) *Protection against spoilage of packages and their contents by micro-organisms, insects, mites and rodents.*
Section 6 (1966) *Temporary protection of metal surfaces against corrosion (during transport and storage).*
Section 7 (1967) *Paper and board. Wrappers, bags and containers.*
Section 8 (1981) *Fibreboard cases.*
Section 9 (1967) *Textile bags, sacks and wrappings.*
Section 10 (1966) *Metal containers.*
Section 11 (1968) *Packaging felt.*
Section 12 (1967) *Cushioning materials (excluding packaging felt).*
Section 13 (1984) *Twines and cords for packaging.*
Section 14 (1966) *Adhesives, closing and sealing tapes.*
Section 15 (1975) *Tensional strapping.*
Section 16 (1968) *Adhesives for packaging.*
Section 17 (1964) *Wicker and veneer baskets.*
Section 18 (1967) *Glass containers and closures.*

Section 19 (1968) *Use of desiccants in packaging.*
Section 20 (1973) *Packaging for air freight excluding livestock* (withdrawn).
Section 21 (1976) *Regenerated cellulose film, plastics film, aluminium foil and flexible laminates.*
Section 22 (1967) *Packaging in plastic containers.*

BS 1224 (1970) *Electroplated coatings of nickel and chromium.*
BS 1282 (1975) *Guide to the choice, use and application of wood preservatives.*
BS 1339 (1965) *Definitions, formulae and constants relating to the humidity of air.*
BS 1344 *Methods of testing vitreous enamel finishes.*
BS 1391 (1952) *Performance tests for protective schemes used in the protection of light gauge steel and wrought iron against corrosion* (withdrawn).
BS 1489 (1972) *Packaging of covered winding wires for electrical purposes.*
BS 1596 (1974) *Fibreboard drums for shipment of goods overseas.*
BS 1615 (1972) *Anodic oxidation coatings on aluminium.*
BS 1706 (1960) *Electroplated coatings of cadmium and zinc on iron and steel.*
BS 1726 *Guide to the design and specification of coil springs.*
BS 1747 Parts 1–7 *Methods for the measurement of air pollution.*
BS 1872 (1984) *Electroplated coatings of tin.*
BS 2000 *Petroleum and its products.*
BS 2011 *Basic environmental testing procedures.*
BS 2087 *Preservation treatments for textiles.*
BS 2540 (1960) *Silica gel for use as a desiccant.*
BS 2541 (1960) *Activated alumina for use as a desiccant.*
BS 2548 (1954) *Wood wool for general packaging purposes.*
BS 2569 *Sprayed metal coatings.*
Part 1 (1964) *Protection of iron and steel by aluminium and zinc against atmospheric corrosion.*
Part 2 (1965) *Protection of iron and steel against corrosion at elevated temperatures.*
BS 2629 *Pallets for materials handling for through transit.*
BS 2644 (1980) *Determination of water absorption of paper and board (Cobb method).*
BS 2782 *Methods of testing plastics.*
BS 2816 (1973) *Electroplated coatings of silver for engineering purposes.*
BS 3130 Parts 1–6 *Glossary of packaging terms.*
BS 3177 (1959) *Method for determining the permeability to water vapour of flexible sheet materials used for packaging.*
BS 3189 (1973) *Phosphate treatment of iron and steel.*
BS 3266 (1981) *Determination of conductivity, pH, water soluble matter, chloride and sulphate in aqueous extracts of textile materials.*
BS 3382 Parts 1–6 *Electroplated coatings on threaded components.*

BS 3433 (1980) *Determination of the water content of paper and board by the oven drying method.*
BS 3452 *Copper/chrome water-borne wood preservatives and their application.*
BS 3453 (1962) *Fluoride/arsenate/chromate dinitrophenol water borne wood preservatives and their application.*
BS 3525 (1962) *Silica gel, cobalt chloride impregnated.*
BS 3597 (1984) *Electroplated coatings of 65/35 tin/nickel alloy.*
BS 3698 (1964) *Calcium plumbate priming paints.*
BS 3745 (1970) *Evaluation of results of accelerated corrosion tests on metallic coatings.*
BS 3810 *Glossary of terms used in materials handling.*
BS 3842 (1965) *Treatment of plywood with preservatives.*
BS 3900 *Methods of test for paints.*
BS 3951 *Freight containers.*
BS 4232 (1967) *Surface finish of blast cleaned steel for painting.*
BS 4290 (1968) *Electroplated coatings of gold and gold alloys.*
BS 4479 (1969) *Recommendations for the flame spraying of ceramic and cermet coatings.*
BS 4601 (1970) *Electroplated coatings of nickel plus chromium on plastics materials.*
BS 4641 (1970) *Electroplated coatings of chromium for engineering purposes.*
BS 4652 (1971) *Metallic zinc-rich priming paint (organic media).*
BS 4672 *Guide to hazards in the transport and storage of packages.*
Part 1 (1971) *Climatic hazards.*
Part 2 (1972) *Climatic hazards (maps and diagrams)*
BS 4758 (1971) *Electroplated coatings of nickel for engineering purposes.*
BS 4921 (1973) *Sherardized coatings on iron and steel articles.*
BS 4950 (1973) *Sprayed and fuzed metal coatings for engineering purposes.*
BS 5056 (1974) *Copper naphthenate wood preservative.*
BS 5073 (1982) *Storage of goods in freight containers.*
BS 5245 (1975) *Phosphoric acid based flux for soft soldered joints in stainless steel.*
BS 5411 (Parts 1–14) *Methods of test for metallic and related coatings.*
BS 5454 (1977) *Recommendations for the storage and exhibition of archival documents.*
BS 5466 (Parts 1–7) *Methods for corrosion testing of metallic coatings.*
BS 5493 (1977) *Code of practice for protective coating of iron and steel structures against corrosion.*
BS 5589 (1978) *Code of practice for preservation of timber.*
BS 5599 (1978) *Hard anodic coatings on aluminium for engineering purposes.*
BS 5666 (Parts 1–19) *Wood preservatives and treated timber.*
BS 5707 (Parts 1–3) *Solutions of wood preservatives in organic solvents.*
BS 5899 (1980) *Method for hydrogen embrittlement test for copper.*

BS 5903 (1980) *Determination of resistance to intergranular corrosion of austenitic stainless steel: copper sulphate–sulphuric acid method (Moneypenny Strauss test).*

BS 6085 (1981) *Method of test for determination of susceptibility of textiles to microbiological deterioration.*

BS 6105 (1981) *Corrosion resistant stainless steel fasteners.*

BS 6137 (1982) *Electroplated coatings of tin/lead alloys.*

BS 6161 (Parts 1–8) *Methods of test for anodic oxidation coatings on aluminium and its alloys.*

BS 6197 (1985) *Preservation and storage of earth moving equipment.*

BS 6338 (1982) *Chromate conversion coatings on electroplated zinc and cadmium coatings.*

BS 6496 (1984) *Powder organic coatings for application and stoving to aluminium alloy extrusions, sheet and preformed sections for external architectural purposes, and for the finish on aluminium alloy extrusions, sheet and preformed sections coated with powder organic coatings.*

BS 6497 (1984) *Powder organic coatings for application and stoving to hot-dipped galvanized hot-rolled steel sections and steel sheet for windows and associated external architectural purposes and for the finish on galvanized steel sections and preformed sheet coated with powder organic coatings.*

BS PD6484 (1979) *Commentary on corrosion at bimetallic contacts and its alleviation.*

BS CP3012 (1972) *Cleaning and preparation of metal surfaces.*

DEFENCE STANDARDS

(Obtainable from the Director of Standardisation, Montrose House, 187 George St, Glasgow G1 1YU.)

DEF STAN 00-1 *Chemical environmental conditions affecting the design of materiel for use by NATO forces in a ground role.*

DEF STAN 00-29 *Fungal contamination affecting the design of military materiel.*

DEF STAN 00-41 *MOD Practices and procedures for reliability and maintainability.*

DEF STAN 00-50 *Guide to chemical environmental contaminants and corrosion affecting the design of military materiel.*

DEF STAN 01-4 *Guide to selection of metallic materials for springs.*

DEF STAN 03-2 *Cleaning and preparation of metal surfaces.*

DEF STAN 03-4 *The pre-treatment and protection of steel parts of specified maximum tensile strength exceeding 1450 N/mm²*

DEF STAN 03-5 *Electroless nickel coating of metals.*

DEF STAN 03-6 *Guide to flame spraying processes.*

Part 1 Unfused metal coatings.

Part 2 Fused metal coatings.

Part 3 Ceramic and cermet coatings.

DEF STAN 03-11 *Phosphate treatment of iron and steel.*
DEF STAN 03-12 *Chromate passivation of brass articles.*
DEF STAN 03-13 *Guide for the prevention of corrosion of metals caused by vapour from organic materials.*
DEF STAN 03-18 *Chromate conversion coatings (chromate filming treatments) for aluminium and aluminium alloys.*
DEF STAN 03-17 *Electrodeposition of gold.*
DEF STAN 03-19 *Electrodeposition of cadmium.*
DEF STAN 03-20 *Electrodeposition of zinc.*
DEF STAN 03-24 *Chromic acid anodizing of aluminium and aluminium alloys.*
DEF STAN 03-25 *Sulphuric acid anodizing of aluminium and aluminium alloys.*
DEF STAN 03-26 *Hard anodizing of aluminium and aluminium alloys.*
DEF STAN 07-55 *Environmental testing of service materiel.*
DEF STAN 13-16 *Paper, kraft, pure.*
DEF STAN 80-15 *Paint, pretreatment primer (etching primer).*

DEFENCE SPECIFICATIONS

(0btainable from HMSO.)

DEF-37 *Phosphoric acid derusting solution (inhibited).*
DEF-38 *The derusting of steel by phosphoric acid solutions.*
DEF-39 *The electrolytic derusting of steel in non-acid solutions.*
DEF-130 *Chromate passivation of cadmium and zinc surfaces.*
DEF-159 *Packaging of electrical indicating instruments.*
DEF-1053 *Standard methods of testing paint, varnish, lacquer and related products.*
DEF-1234A *Production requirements for service packaging.*
DEF-1237 *Wrapping, mouldable, waxed, grease resisting.*
DEF-1244 *Packaging of engineers cutting tools.*
DEF-1251 *Paper, tissue, wrapping.*
DEF-1278 *Packaging of plugs and sockets.*
DEF-1316 *Paper, wrapping, grease resistant.*

DEFENCE GUIDES

(Obtainable from the Director of Standardisation, Montrose House, 187 George St, Glasgow G1 1YU.)

DG-8 *Defence Guide: treatments for the protection of metal parts of service stores and equipments against corrosion.*
DG-10 *Metallic springs: protection against corrosion.*
DG-12A *Petroleum oils and lubricants (POL) and allied products.*
DG-13 *Design aspects of chromium plating for engineering purposes.*

DTD SPECIFICATIONS

(Obtainable from HMSO.)

DTD 599A *Non-corrosive flux for soft soldering.*
DTD 911C *Protection of magnesium alloys against corrosion.*
DTD 935 *Surface sealing of magnesium rich alloys.*
DTD 938 *Gold plating.*
DTD 940 *The cadmium coating of very strong steel parts by vacuum evaporation.*
DTD 941 *Surface coating by use of detonation, flame and plasma spraying processes.*
DTD 942 *Anodizing of titanium and titanium alloys.*

DEPARTMENT OF INDUSTRY GUIDES AND ADVISORY DOCUMENTS

(Obtainable from the National Corrosion Service, National Physical Laboratory, Teddington, Middlesex, TW11 0LW.)

Guides to Practice in Corrosion Control

No. 1 (1979) *Sources of corrosion information.*
No. 2 (1979) *Corrosion of metals by wood.*
No. 3 (1979) *Packaging for handling and transport of coated goods in the construction industry.*
No. 4 (1980) *Stress corrosion.*
No. 5 (1981) *The handling and storage of coated wrapped steel pipes.*
No. 6 (1981) *Temporary protection.*
No. 7 (1981) *The corrosion of steel and its monitoring in concrete.*
No. 11 (1982) *Avoidance of corrosion during chemical cleaning of plant.*
No. 12 (1982) *Paint for the protection of structural steelwork.*
No. 13 (1982) *Surface preparation for painting.*
No. 14 (1982) *Bimetallic corrosion.*
No. 15 (1982) *Corrosion in agriculture.*

Controlling Corrosion (Committee on Corrosion)

No. 1 (1980) *Methods.*
No. 2 (1980) *Advisory services.*
No. 3 (1980) *Economics.*
No. 4 (1983) *Specifications and standards.*
No. 5 (1979) *Case studies.*
No. 6 (1984) *Controlling corrosion.*

Index